EXPLORING PRECALCULUS

with

DANIEL SCHER
SCOTT STEKETEE
PAUL KUNKEL
IRINA LYUBLINSKAYA

Key Curriculum Press
Innovators in Mathematics Education

Project Editor:	Scott Steketee
Consulting Editor:	Christopher David
Editorial Assistant:	Shannon Miller
Reviewer:	Dan Lufkin
Sketch Checkers:	Danny Zhu and Aaron Madrigal
Production Director:	McKinley Williams
Production Editor:	Christine Osborne
Copyeditor:	Erin Milnes
Production Coordinator, Text Designer, and Compositor:	Ann Rothenbuhler
Art Editor:	Jason Luz
Art and Design Coordinator:	Kavitha Becker
Cover Photo Credits:	G. K. and Vikki Hart/Brand X Pictures/ PictureQuest
Prepress and Printer:	Data Reproductions

Executive Editor: Casey FitzSimons
Publisher: Steven Rasmussen

Exploring Precalculus Sketches CD-ROM
Key Curriculum Press guarantees that the *Exploring Precalculus* Sketches CD-ROM that accompanies this book is free of defects in materials and workmanship. A defective CD-ROM will be replaced free of charge if returned within 90 days of the purchase date. After 90 days, there is a $10.00 replacement fee.

Key Curriculum Press
1150 65th Street
Emeryville, CA 94608
510-595-7000
editorial@keypress.com
www.keypress.com

10 9 8 7 6 5 4 3 2 1   09 08 07 06 05

ISBN 1-55953-680-2

CONTENTS

CD-ROM CONTENTS

The list below was current when this book went to press, but be sure to check the Read Me.gsp file on the actual CD-ROM for additional activities, sketches, and tools.

ACTIVITY SKETCHES

The CD-ROM has a folder for every chapter in this book, containing all the sketches required for the activities in that chapter. For many of the activities there are also presentation sketches that you can use to present the activity as a classroom demonstration.

SUPPLEMENTAL ACTIVITIES

This folder on the CD-ROM contains several complete activities, including sketches, handouts, and activity notes. The handouts are in PDF format to make it easy to print them or to give them to students electronically. (To read PDF files, use Adobe Reader, a free download from www.adobe.com/products/acrobat/readermain.html.)

Spirals

Students manipulate an equation in polar coordinates to form Archimedes' spiral, Fermat's spiral, and a logarithmic spiral, in both continuous form and discrete form.

Cycloid

Students use vector addition and parametric functions to define a cycloid, and modify the equations to create variations, such as the locus of a point on a circle rolling up an incline.

Epicycloid and Hypocycloid

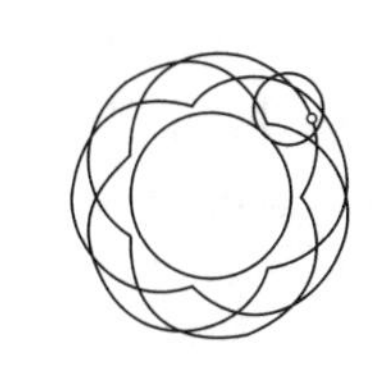

Students apply the concept of vector addition to define parametric functions for an epicycloid and a hypocycloid.

Phasor Diagrams

Phasors are an analog of vectors, used by electrical engineers to understand the combinations of sinusoids that make up alternating current. Students manipulate phasors to calculate sums of sinusoids.

Frame of Reference

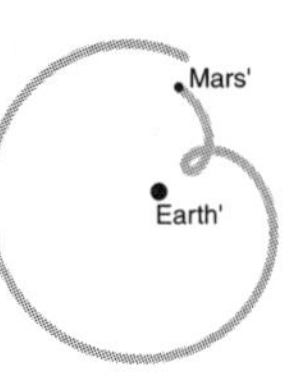

Students modify the frame of reference in order to simplify several problems: a moving-car problem, a clock-hand problem, and a retrograde-motion problem. They solve the problems using moving and rotating frames of reference.

Chaos

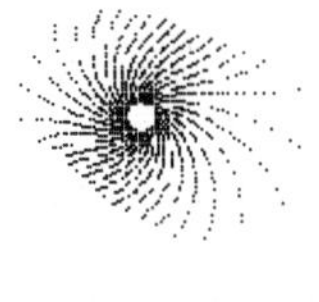

Students create various fractal designs (including the Sierpiński triangle and Mira, Julia, and Mandelbrot fractals) as strange attractors.

The Mandelbrot Set

Students investigate the Julia mapping and see how it is used to generate the Mandelbrot set. The accompanying sketch produces a beautiful colored image of the Mandelbrot set and allows the student to zoom in on any portion of the set.

Barnsley's Fern

Students plot points using four pairs of iterated functions. By choosing randomly among the function pairs, they create a fractal Barnsley's Fern.

SUPPLEMENTAL SKETCHES

This folder on the CD-ROM contains sketches that you can use for demonstrations, or that are useful as templates for creating various graphs.

Piano.gsp

In this sketch, students press the piano keys to create chords or dissonances and see the shapes of the resulting sound waves.

3D Graph Template.gsp

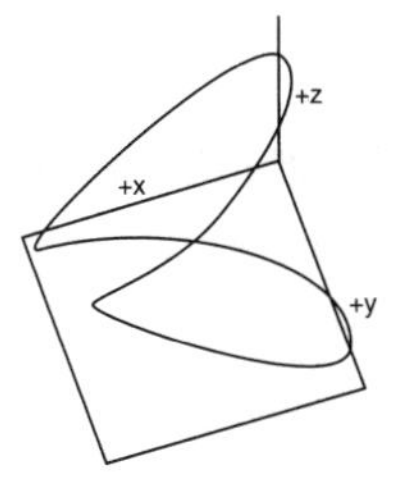

Students can use this sketch to plot points and curves in three dimensions, and then spin the three-dimensional axes to view their creations from different angles.

Trig Coords.gsp

Students can use the coordinate system in this sketch to plot functions using radians, expressed as fractions of π, on the x-axis.

Zooming Coordinate System.gsp

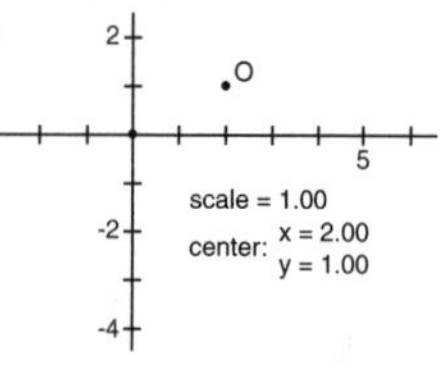

Students can use this coordinate system to plot functions and then zoom in or out on the resulting graphs.

Light Speed.gsp

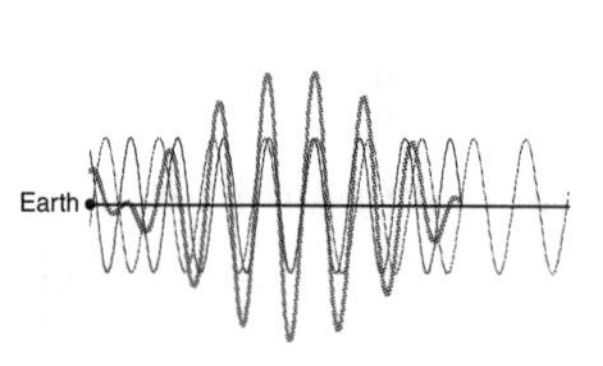

This sketch illustrates a wave packet traveling faster than the speed of light and shows how such a packet cannot overcome the limitations of Einstein's Theory of Special Relativity.

SUPPLEMENTAL TOOLS

This folder on the CD-ROM contains tools that are useful for many purposes: intersecting functions, working with vectors and matrices, and so forth.

0-2pi Radian Measure.gsp

These tools allow students to measure angles in radians, expressed as fractions of π or as decimal values between 0 and 2π.

3D Tools.gsp

Students use these tools to create 3D coordinate systems and plot points on them using rectangular, cylindrical, or spherical coordinates.

Conic Plus.gsp

Students use these tools to construct geometric loci of parabolas, ellipses, and hyperbolas using a variety of methods.

Conic Intersections.gsp

Students use these tools to find the intersections of straight objects with parabolas, ellipses, or hyperbolas defined as loci.

Intersect Functions.gsp

These tools find the first zero of a function, or the first intersection of two functions, within a specified domain.

Matrix Brackets.gsp

This appearance tool creates adjustable matrix brackets that can be sized and positioned to enclose a rectangular array of parameters.

Matrix Tools.gsp

Students can multiply a matrix by a vector or another matrix and find the determinant, adjoint, or inverse of a matrix. The sketch contains tools for 2-, 3-, 4-, and 5-dimensional matrices.

Vector Tools.gsp

Students can find vector sums, lengths, and dot products in 2, 3, 4, and 5 dimensions and vector cross products in 3 dimensions.

QUESTIONS ABOUT USING THIS BOOK

I don't have regular access to a computer lab. How can I use the activities?

Most of the activities can also be used as classroom demos, with a single computer and an LCD projector. We've provided many presentation sketches for this purpose. Alternatively, if you have a computer in your room, you can assign activities to particular students as enrichment work.

How can I use the CD-ROM?

You are entitled to make the contents of the CD-ROM accessible on the computers that your students use, provided that the computers are covered by a license for Sketchpad. (Sketchpad itself is a separate purchase.) You can copy the contents of the CD-ROM to the hard drives of these computers or, if you have a Sketchpad site license, you can put the contents of the CD-ROM on a network server.

What's on the CD-ROM?

The CD-ROM contains the activity and presentation sketches for the printed activities, organized into folders by chapter. It also contains a number of supplemental activities, ketches, and tools. The supplemental activities come with tudent handouts, sketches, and activity notes. (You will eed Adobe Reader to view and print the handouts and otes. This free download is available at www.adobe.com/ roducts/acrobat/readermain.html.) The supplemental ketches and tools include things like a sketch that makes it asy to create your own three-dimensional graphs and a tool hat finds the intersection of two functions. The complete sting is on the CD-ROM itself and also in the CD-ROM Contents immediately following the Contents in this book.

What's the easiest way for me to tell if students are answering he questions correctly?

ook at the Activity Notes in the back of the book. Each ctivity has a corresponding Activity Note that provides seful commentary and the answer to every question in he activity.

Some of the activities are on topics that are not included in ny course. Why is this?

Every precalculus textbook covers different topics, in a different order. The activities in this book address many f those topics, probably including some that are not in our particular text. You should still find plenty of ctivities that do address topics in your book, and even hose that don't can be useful as enrichment activities.

How can I get a summary of all the activities?

The CD-ROM contains two files (**Contents.doc,** in Microsoft Word format, and **Contents.pdf,** in Adobe Reader format) that list each activity, tool, and template, with its required Sketchpad proficiency and a brief description. If the CD-ROM is not handy, you can browse he Activity Notes at the end of the book; these notes also ontain the description and proficiency information.

My students have never used Sketchpad before. What hould I do?

The first activity (Translation of Functions) is especially designed to introduce students to using Sketchpad, without ompromising the activity's value for students already amiliar with the program. In addition, many of the ctivities are based on prepared sketches and do not require any Sketchpad experience. Look at the "Sketchpad Proficiency" paragraph in the Activity Notes (or the Proficiency column in the Contents file on the CD-ROM) to determine the level of experience required for each activity.

QUESTIONS ABOUT PRECALCULUS ACTIVITIES

How can transformations of functions be related to the geometric transformations with which my students are already familiar?

In the activities in Chapter 1, students explore both geometric and symbolic function transformations and discover the relationships between these two ways of transforming functions.

How can students define the sine function based on a unit circle?

In Trigonometry Tracers (Chapter 2), students animate a point around a circle, measure its y-coordinate, and use that measurement to create a graph of the sine function.

Students visualize the Pythagorean Theorem in terms of squares on the sides of a right triangle. Is there a similar visualization for the Law of Cosines?

In The Law of Cosines (Chapter 3), students build a Sketchpad model that relates the terms of $c^2 = a^2 + b^2 - 2ab(\cos C)$ to the areas of three squares and two parallelograms.

Graphing a logarithm function is a snap with a graphing calculator. How can a Sketchpad activity enliven this topic?

In A Sequence Approach to Logs (Chapter 4), students stitch together a log curve by plotting a geometric sequence against an arithmetic sequence. Subsequent logarithm activities introduce semilog and log-log graphs.

What role can Sketchpad play in the study of statistics and probability?

In Linear Regression (Chapter 5), students drag data points and observe the simultaneous effect on the regression line. In Wait for a Date, a simulation suggests a geometric approach to computing the probability of two people meeting within a specified time period.

How can students use matrices to perform geometric transformations like rotation and dilation?

In Matrix Transformations (Chapter 6), students use 2×2 matrices to rotate, dilate, reflect, and stretch shapes in the plane.

How can I make the study of complex numbers more tangible?

In Multiplication of Complex Numbers (Chapter 7), students use triangle similarity to frame an understanding of multiplication in the complex plane. In A Geometric Approach to $e^{i\pi}$, students develop an intuitive sense of why $e^{i\pi}$ equals -1 by examining the limit definition of e^x from a geometric standpoint.

What is there that's "geometric" about a geometric series?

In "Area Models of Geometric Series" and "A Geometric Series Staircase" (Chapter 8) students create attractive geometric models of such series.

How can I incorporate fundamental ideas from calculus into a precalculus course?

The seven activities in Chapter 9 provide dynamic visual introductions to fundamental topics such as limits, rate of change, and accumulation.

To the Student

Beauty is the first test;
there is no permanent place in the world
for ugly mathematics.

G. H. Hardy, in *A Mathematician's Apology*

It's our hope that you will enjoy exploring the activities in this book, and that one or more of them will bring you a bit of the joy associated with what has been called the "Aha!" moment. We have taken pleasure in writing the activities, both because we find the mathematics beautiful and because we would like to share that appreciation with you. The visual, dynamic nature of these activities will, we believe, allow you to discover some novel mathematical connections and to appreciate the subtlety and power of trigonometry, complex numbers, and other precalculus topics.

We are convinced that by investigating and manipulating the mathematics yourself, by observing on the computer screen the effects of your actions, and by thinking deeply about the results, you will take greater satisfaction in what you have learned. You may even be inspired to continue your explorations beyond the activities as we've written them! Tell us about your investigations by sending an email to exploring_precalc@keypress.com.

Daniel Scher
Scott Steketee
Paul Kunkel
Irina Lyublinskaya

To the Teacher

We created the activities in this book because we were excited about them. We didn't base our choices on a specific curriculum, though we have covered the majority of topics found in current precalculus texts. Such books differ greatly in subject matter and even in where they draw the lines between advanced algebra, precalculus, and calculus. Regardless of the content of your particular course, you'll find plenty of activities that address topics you do teach, and that will make those topics clearer and more compelling to your students.

Whenever possible, approach these activities with students in a hands-on way; students learn best and take ownership of their own learning by digging headfirst into the mathematics. Pay particular attention to the accompanying questions, which are designed to encourage students to make conjectures and to think about what they are doing and observing. The worth of the activity is determined by the student's intellectual involvement in the investigation, and answering the questions thoughtfully fosters such involvement.

Similarly, encourage students to discuss the mathematics among themselves. Often activities are more successful when students work in pairs and discuss their conjectures with their partners. One of the main purposes of activities such as these is to generate mathematical conversations in your classroom.

We hope that both you and your students will benefit from and enjoy the dynamic visualization of precalculus topics that serves as the basis for these activities. We'd love to hear about your experiences and suggestions; send an email to exploring_precalc@keypress.com.

Best wishes to you and your students!

Daniel, Scott, Paul, and Irina

Acknowledgments

Thanks to Nick Jackiw for originally creating Sketchpad and for guiding its development into such an exciting tool for the dynamic visualization of mathematics.

Thanks also to Paul Foerster. Several of the activities in Chapter 9 are based on the first chapter of his textbook *Calculus: Concepts and Applications.*

Daniel thanks Steven Chanan for pursuing the idea of a Sketchpad precalculus module well before the project was a certainty. As the book got under way, Scott, Paul, and Irina contributed enormously to its content and shape, proving the value of long-distance collaboration (New York to California to China). Finally, much closer to home, Sara provided support during the entire writing process.

Scott thanks Shannon Miller for keeping the project on track by taking care of so many project details, Chris David for being ready on a moment's notice with a sympathetic ear and good advice, Steven Chanan for contributing several excellent activities, and Nan Langen Steketee for her extraordinary patience and support when he brought the work on this project home, both physically and mentally. Finally, he expresses his deep gratitude to Daniel, Paul, and Irina, without whose labor of love this book would not exist.

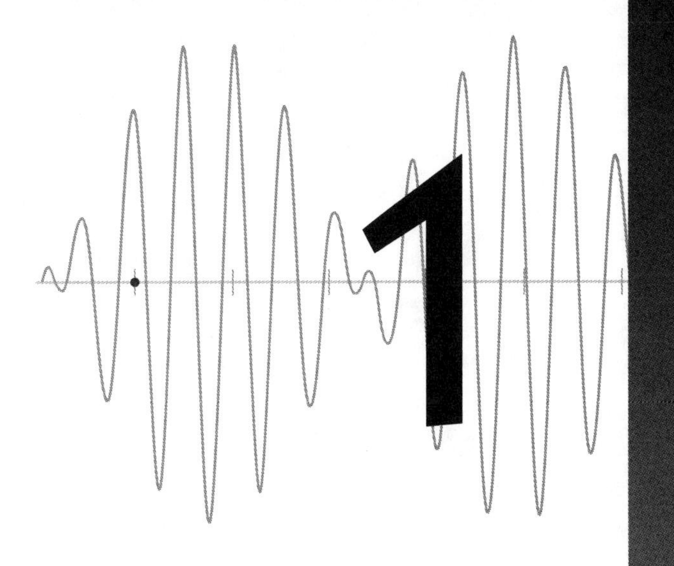

Function Transformations

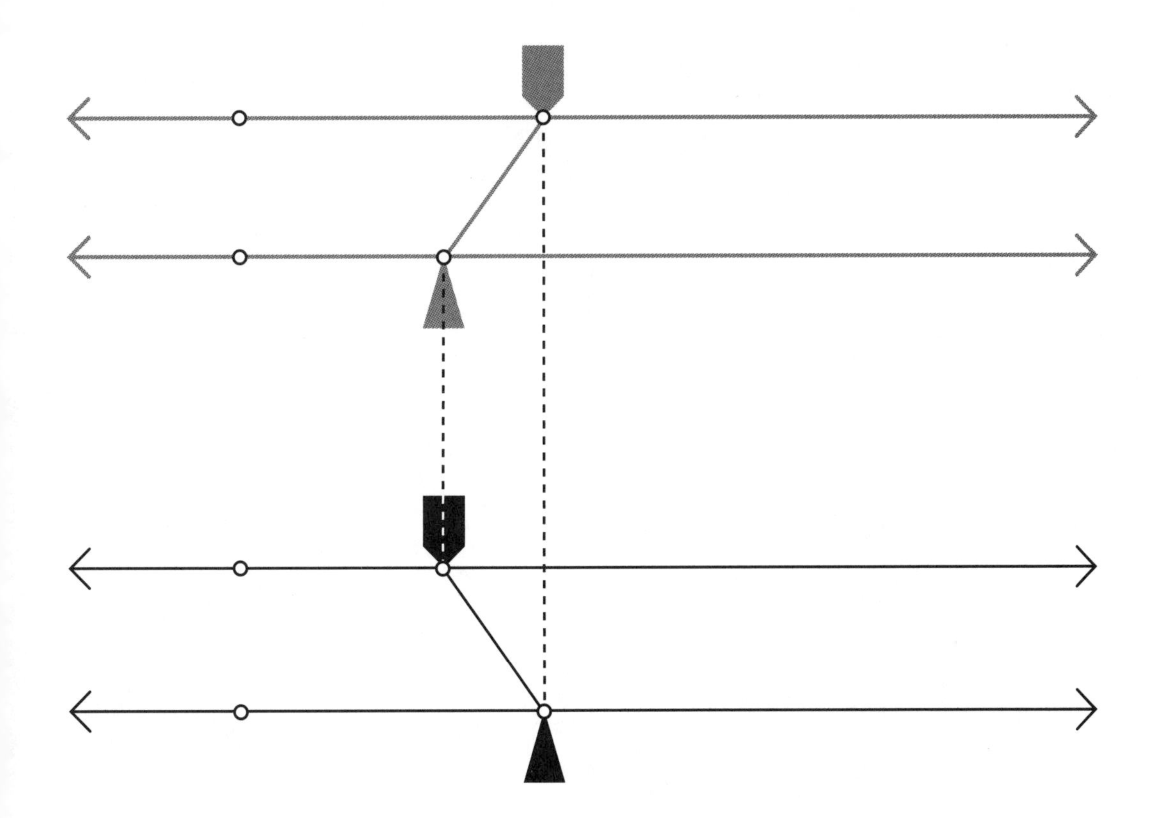

Translation of Functions

A mathematical function is a set of ordered pairs (x, y) in which any value of x in the domain is associated with only one value of y in the range.

Many ordinary measuring devices are based on mathematical functions. For instance, the odometer of a car is operated by the rotating wheels, so for any given number of rotations after the car leaves the factory (the x-value), there is only one value that appears on the odometer dial (the y-value). In other words, there is a function that relates the number of revolutions of the wheels to the number of miles (or kilometers) that appears on the dashboard's odometer.

In addition to the odometer, many cars have a trip counter that allows you to measure the miles traveled on a specific trip. The numbers on the trip counter differ from the numbers on the odometer, and the odometer and the trip counter correspond to two different mathematical functions. Both are based on the number of revolutions of the wheels, and they behave similarly—if the odometer increases by 249 miles, so does the trip counter, provided you don't reset the trip counter during the drive. The trip counter's function is called a translation of the odometer's function.

In this activity you will explore the mathematical relationship between an original function and another function that is a translation of the original.

EXPLORE TRANSLATIONS OF FUNCTIONS

Just as in geometry, there are other kinds of function transformations besides translation. You'll learn about other transformations in later activities.

In geometry, a transformation that moves an object to a new location without changing the shape, size, or orientation of the object is called a *translation.* Similarly, a function transformation that moves the plot of the function to a new location without changing the shape, size, or orientation is also called a *translation.*

The diagram at right shows an original pre-image function $y = f(x)$, a translation vector OV, and a translated image function $y = t(x)$.

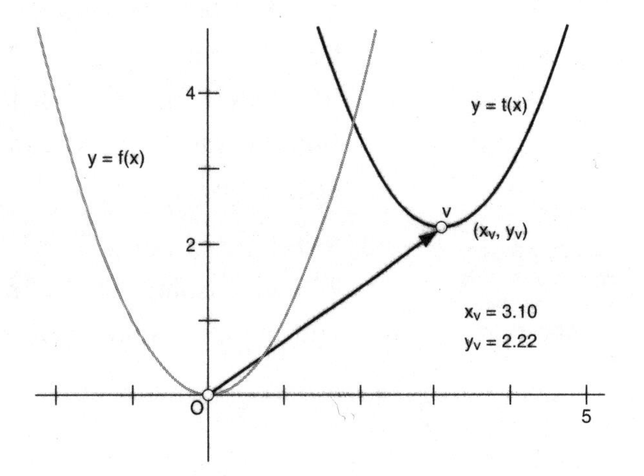

If this is your first Sketchpad activity, be sure to use Sketchpad's Help menu to get any help you need to use the menus and the tools.

Q1 How do you think the graph of the function $y = t(x)$ will change for different values of x_V and y_V? Will it ever get narrower or wider? What will happen when x_V increases or decreases? What will happen when y_V changes?

If you're an experienced Sketchpad user, follow the instructions in each of the following summaries to investigate translations of functions. If you need more guidance, first read each summary, and then use the detailed instructions that follow the summary.

TRANSLATE A POINT ON THE FUNCTION PLOT

In this section you'll plot a function, define a translation vector, construct a point on the function plot, and translate the point by the vector.

Summary

If you are experienced with Sketchpad, you can follow the Summary directions and skip the numbered steps that follow.

Open the sketch **Translation.gsp** from the **1 Function Transformations** folder. The Activity page contains a coordinate system and a vector, *OV*. Plot the function $f(x) = -x^3 - x^2 + 2x$, and make the plot thick and blue. The plot is the pre-image; you will construct a translated image.

Translate point *P* so that you can later translate the entire function plot.

Mark vector *OV* as Sketchpad's vector of translation. Construct a point *P* anywhere on the function plot, and translate point *P* by vector *OV*. Label the translated image *P'*.

Detailed Instructions

1. Open the sketch **Translate.gsp** from the **1 Function Transformations** folder.

Click the ^ key before typing an exponent.

2. To plot the function $f(x) = -x^3 - x^2 + 2x$, choose **Graph | Plot New Function.** In the New Function Calculator that appears, click the keys on the keypad to enter the function. Click OK to plot the function. Both the function $f(x)$ and the plot appear, and both are selected.

To select only the plot, click in empty space to deselect everything, and then click the plot to select it.

3. Select the plot only. This is the original pre-image. Choose **Display | Line Width | Thick** to make it thick, and choose **Display | Color** to make it blue.

4. To mark the vector from O to V as Sketchpad's vector of translation, select O and V in order, and choose **Transform | Mark Vector.**

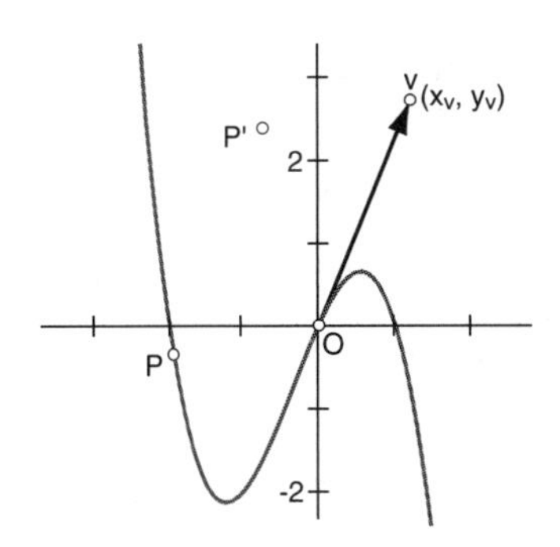

Click the **Text** tool on the new point to show its label. Then double-click the **Text** tool on the label (not the point) to change it to P.

5. Use the **Point** tool to construct a point anywhere on the function plot. Label it P.

6. Select point P and choose **Transform | Translate** to translate it by vector OV.

TRANSFORM THE FUNCTION PLOT GEOMETRICALLY AND ALGEBRAICALLY

In this section you'll observe the behavior of P' and use it to construct geometrically an image of $f(x)$. You'll also plot an algebraic image: a function based on $f(x)$.

Q2 Drag point P along the function plot, and observe the behavior of point P'. Write down your observations.

Summary

Construct the locus of P' as P moves along the plot of $f(x)$, make it thick and red, and label it $t(x)$. This is the image of the $f(x)$ plot translated by vector OV. Move point V, and observe the effect of the translation vector on the $t(x)$ plot.

Next create a new function by using addition to transform $f(x)$ algebraically. To do so, press the *Show Sliders* button to display two sliders. Use the value of the sliders to plot $g(x) = f(x + a) + b$. Move each slider to observe its effect on the $g(x)$ function plot.

Detailed Instructions

Use **Display | Line Width** and **Display | Color** to change the thickness and color of the locus.

Use the **Text** tool to label the locus.

7. Construct the locus of P' as P moves along the function plot by first selecting both P and P' and then choosing **Construct | Locus.** This is the plot of $t(x)$, the translated image of $f(x)$. Make this image dashed and black. Label it $t(x)$.

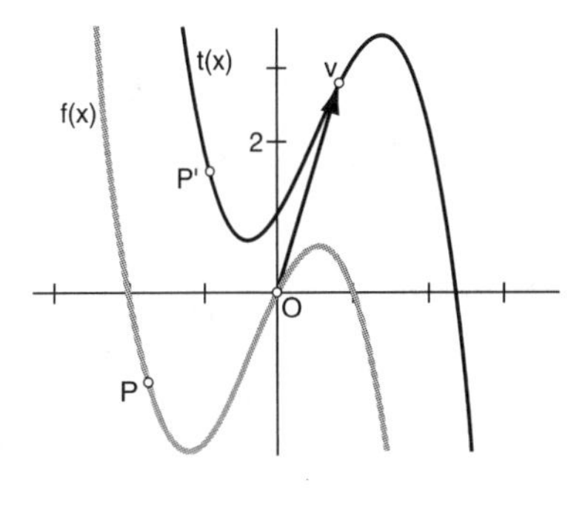

8. Drag point P and observe point P'. Move point V, and observe the effect of the translation vector on the translated image $t(x)$.

Now try to find an algebraic equation that corresponds to the translated function $t(x)$.

Try dragging a slider to change its value.

9. Press the *Show Sliders* button. Two sliders appear.

You may have to move the Calculator to the side if it's covering the function. Be sure to click the function, not the function plot.

10. Using the values of the sliders, plot the function $g(x) = f(x + a) + b$. To do so, choose **Graph | Plot New Function.** The New Function Calculator appears. In the sketch, click on the function $f(x) = -x^3 - x^2 + 2x$. The Calculator display shows $f(\,|\,)$. Next click the **x** and the + keys on the Calculator's keypad and click on the value of the *a* slider in the sketch. Click the closing parenthesis ")" on the Calculator and then the + key. Finally, click on the value of the *b* slider in the sketch.

The new function is labeled $g(x)$.

11. When the function looks correct, click OK. Make it thick and light green.

12. Move each slider back and forth and observe the effect on the plot.

MATCH THE TRANSFORMATIONS

In this section you will manipulate both transformed images you've created and describe their behavior. Then you will edit the algebraic transformation $g(x)$ so that it always matches the geometric transformation $t(x)$.

Q3 As you drag *V*, does the geometrically constructed image—the $t(x)$ plot created using the **Locus** command—always have the same shape, size, and orientation as the pre-image $f(x)$? Does this mean it is a translated image of the $f(x)$ plot?

Q4 As you move the sliders, does the algebraically constructed image—the plot of $g(x)$—always have the same shape, size, and orientation as the pre-image $f(x)$? Does this mean it is a translated image of the $f(x)$ plot?

You can click the buttons provided to set *a* or *b* to zero.

Q5 What happens to the graph of $g(x)$ when you set *b* to zero and change the value of *a*? Specifically, what happens if *a* is negative? Positive? Why does this occur?

Q6 What happens to the graph of $g(x)$ when you set *a* to zero and change the value of *b*? Specifically, what happens if *b* is negative? Positive? Why does this occur?

To create a table, select the four values *a*, *b*, x_V, and y_V and choose **Graph | Tabulate.**

Q7 Set *a* to 3 and *b* to zero. Manipulate the vector *OV* to make $t(x)$ match $g(x)$. In what direction must *OV* point? Start a table with the values of *a*, *b*, x_V, and y_V.

To add values to a table, double-click the table with the **Arrow** tool.

Q8 Set *a* to zero and *b* to -2. Manipulate the vector *OV* to make $t(x)$ match $g(x)$. In what direction must *OV* point? Add these new values of *a*, *b*, x_V, and y_V to your table.

CONCLUSION

Q9 Drag *V* so that *OV* is neither horizontal nor vertical. Manipulate the sliders to make $g(x)$ match the transformed image $t(x)$. What values must you use? Add these values of *a*, *b*, x_V, and y_V to your table.

Q10 Drag the sliders to change a and b, and then drag V so that the functions match once more. Add these values of a, b, x_V, and y_V to your table.

Q11 What pattern do you see in the four sets of values you have recorded?

Click on the values of x_V and y_V in the sketch to enter them into the Calculator.

13. Once you determine the pattern, double-click the **Arrow** tool on the function definition of $g(x)$, and change the function definition so that it uses the values of x_V and y_V rather than a and b.

Q12 Test your result by dragging V. Do the two functions continue to match, no matter where you drag V?

EXPLORE MORE

Q13 What happens if you change the original function $f(x)$? Is the equation that you used for $g(x)$ still correct? To check your answer, edit function $f(x)$ and change it to a totally different function—for instance, an exponential function like $f(x) = 2^x$ or a trig function like $f(x) = 3 \cdot \sin hr\ (2x)$. Try several different functions for $f(x)$ to verify that the transformed function $g(x)$ represents the general result. Write down each of the functions you tried for $f(x)$.

Q14 Can you find a function that you can translate to the same image by two different translations? Use the Challenge page of the sketch to explore this question.

Dilation of Functions

When you weigh yourself, you may use a bathroom scale with a rotary dial. For such a scale, there's a mathematical function that relates the weight on the scale to the rotation of the dial (and thus to the number you read from the dial). In the United States the bathroom scale will give you your weight in pounds. Some bathroom scales also show your weight in kilograms. On these scales, you will read a different numerical value for your weight in kilograms than for your weight in pounds. Both of these functions relate the same weight to the same rotation of the dial, so this new function is a transformation of the original function. You could describe the kilogram-weight function as a multiplication of the pounds-weight function by a constant factor. Such a transformation is called a stretch or compression, because one set of numbers on the dial appears to be stretched or compressed relative to the other.

In this activity you will determine the mathematical relationship between an original pre-image function and another function that is its stretched or compressed image.

EXPLORE STRETCHING AND COMPRESSING

In geometry, a transformation that changes the size of an object without changing its shape or orientation is called a *dilation*. Similarly, an algebraic transformation that changes the size of a function plot is sometimes called a *dilation*.

This activity uses the terms *dilate, stretch,* and *compress.* You may also see the terms *expand, contract,* and *magnify* used to describe such transformations.

You can dilate functions independently in the x (horizontal) or y (vertical) directions. This makes algebraic dilation different from geometric dilation, so the terms *stretch* and *compress* are often used instead of *dilate* when referring to functions.

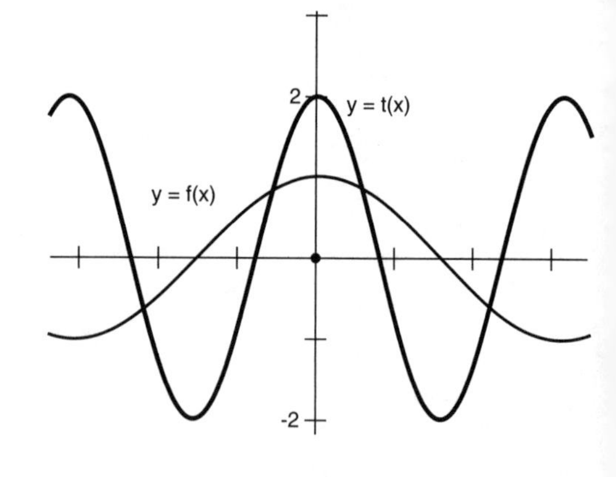

The diagram at right shows an original pre-image function $y = f(x)$ and a dilated image $y = t(x)$.

Q1 Describe the changes you see in the graph of the image function plot $y = t(x)$ compared to the pre-image function plot $y = f(x)$. Has the image been stretched or compressed horizontally? Vertically? By how much?

PLOT A POINT WITH DILATED COORDINATES

In this section you will plot a function, construct a point on the function plot, dilate its x- and y-coordinates, and construct an image point using the dilated coordinates.

With the plot of the function selected, choose **Display | Line Width | Thick,** to make it thick. Choose **Display | Color** to make it blue. Choose **Display | Show Label** to show the label.

1. Open the document **Dilation.gsp.** The Activity page contains a coordinate system and a *Show Sliders* button. Choose **Graph | Plot New Function,** and enter the function $f(x) = -0.5x^2 + 4$. Make the plot of the function thick and blue. Show the label of the function plot.

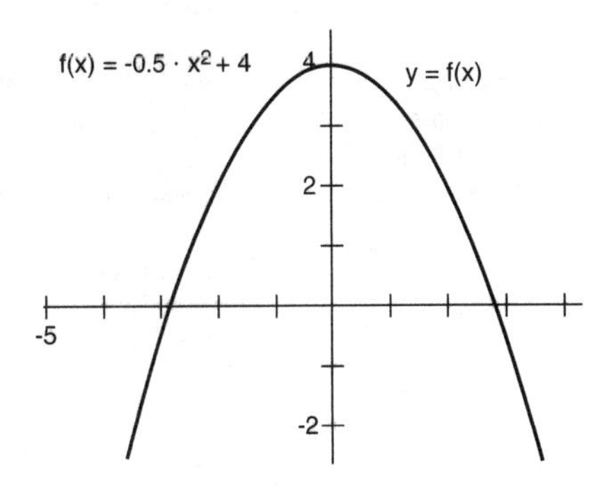

2. Use the **Point** tool to construct a point anywhere on the function plot. Use the **Text** tool to label the point P.

3. Measure the x-coordinate of point P by selecting the point and choosing **Measure | Abscissa (x).** Similarly, measure the y-coordinate.

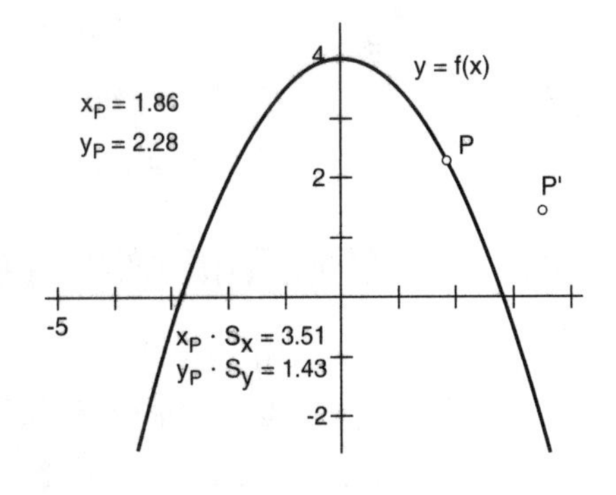

Choose **Measure | Calculate** to use the Calculator. Click on the measurements in the sketch to enter the values of x_P and S_x.

4. Press the *Show Sliders* button to display four sliders. You will use the first two (S_x and S_y) as scale factors in the x and y directions.

5. Use Sketchpad's Calculator to calculate the value of $x_P \cdot S_x$. This is the dilated value of x.

6. Similarly, calculate $y_P \cdot S_y$, the dilated y-value.

7. To construct an image point with dilated coordinates, select measurements $x_P \cdot S_x$ and $y_P \cdot S_y$ in order, and choose **Graph | Plot As (x,y).** Label this point P'.

TRANSFORM THE FUNCTION PLOT GEOMETRICALLY AND ALGEBRAICALLY

In this section you will observe the behavior of P' and use it to construct geometrically an image of $f(x)$. You will also plot an algebraic image: that is, a function $t(x)$ that is based on the function $f(x)$.

To turn on tracing, select the point and choose **Display | Trace Point.**

8. Turn on tracing for point P'.

Q2 Drag the sliders to set $S_x = 2$ and $S_y = 1$. Drag point P along the function plot, and observe the behavior of point P'. Write down your observations.

To erase traces, choose **Display | Erase Traces.**

Q3 Set $S_x = 1$ and $S_y = 0.5$. Erase the previous trace, and drag point *P* along the function plot. Observe the behavior of point *P′*. Write down your observations.

Q4 Now set $S_x = 0.5$ and $S_y = 2$. Erase the previous trace, drag point *P*, and again describe the result.

To turn off tracing, select the point and choose **Display | Trace Point.** (This removes the check mark in the menu.)

9. Turn off tracing for *P′*. Construct the locus of *P′* as *P* moves along the function plot of $f(x)$ by selecting both *P* and *P′* and then choosing **Construct | Locus.** Label this image $t(x)$, and make it dashed and black.

10. Drag the sliders S_x and S_y, and observe the effect of scale factors on the stretched image $t(x)$.

Your objective now is to find an algebraic equation that corresponds to the image function $t(x)$. You will create a new function by using multiplication to transform $f(x)$ algebraically.

To enter *c* or *d* into the Calculator, click the value in the sketch. To enter *f* into the Calculator, click the function (not the plot) in the sketch.

11. Plot the function $g(x) = c \cdot f(d \cdot x)$ by choosing **Graph | Plot New Function.** Label this function *g*, and make its plot thick and green.

12. Move sliders *c* and *d* back and forth, and observe the effect of the slider's values on the plot of *g*.

MATCH GEOMETRIC AND ALGEBRAIC TRANSFORMATIONS

In this section you will manipulate both transformed images you have created and describe their behavior. Then you will edit the algebraic transformation $g(x)$ so that it always matches the geometric transformation $t(x)$.

Q5 What happens to the graph of $g(x)$ when you set $d = 1$ and change the value of *c*? Specifically, what happens if $c > 1$? $c < 1$? Why does this occur?

Q6 What happens to the graph of $g(x)$ when you set $c = 1$ and change the value of *d*? Specifically, what happens if $d > 1$? $d < 1$? Why does this occur?

To tabulate values, select the values that should appear in the table and choose **Graph | Tabulate.**

Q7 Set $c = 2$ and $d = 1$. Can you manipulate S_x and S_y to make $t(x)$ match $g(x)$? Tabulate the values of S_x, S_y, *c*, and *d*.

Q8 Set $c = 1$ and $d = 0.5$. Can you manipulate S_x and S_y to make $t(x)$ match $g(x)$? Double-click the table to add these values of S_x, S_y, *c*, and *d* to the table.

Q9 Set the values of S_x and S_y so that neither value is equal to 1. Can you manipulate the sliders *c* and *d* to make $g(x)$ match the transformed image $t(x)$? What values of *c* and *d* must you use? Add these values of S_x, S_y, *c*, and *d* to the table.

Q10 Change the values of S_x and S_y once again, and adjust the values of *c* and *d* to match functions $t(x)$ and $g(x)$. Add these values of S_x, S_y, *c*, and *d* to the table.

Q11 What pattern do you see in the four sets of values in your table?

To edit a function, double-click it with the **Arrow** tool.

13. Once you determine the pattern, edit the function $g(x)$, changing it to use the values of S_x and S_y rather than c and d.

14. Test your results by dragging sliders S_x and S_y. Do the two functions continue to match no matter what values are used for S_x and S_y?

Q12 What if you change the original function $f(x)$? Will the equation that you used for the transformed function $g(x)$ still be correct? To check your answer, edit the function $f(x)$ and change it to a totally different function—for example, $f(x) = 3 \cdot \cos(2x)$. Try several different functions for $f(x)$ to verify that the transformed function $g(x)$ represents the general result. Write down each of the functions you used for $f(x)$.

EXPLORE MORE

Now you can combine the results of your explorations of function translations, stretches, and compressions.

Q13 If a pre-image function $f(x)$ is translated by a vector (x_V, y_V) and then scaled by scale factors S_x and S_y, what is the equation of the transformed image function $h(x)$ in terms of x_V, y_V, S_x, and S_y?

Q14 If the pre-image is scaled first and then translated, is the equation of the transformed image function the same as in Q13? If not, what is it?

On the last three pages of the **Dilation.gsp** document you will find three challenges. Each challenge includes the plot of a function $f(x)$ in blue, the plot of the function transformation $g(x)$ in red, and a *Show/Hide Sliders* button. Your goal is to determine an algebraic equation for the given function transformation. Follow the directions on each Challenge page of the document.

Q15 For each challenge, first make a conjecture describing the required transformation. (For instance, "You must translate $f(x)$ by $(-1, 2)$ and dilate it by $(0.5, 1.0)$ to match $g(x)$.") Then construct a transformed function $h(x)$, and record slider positions that actually correspond to the required transformation. How close was your conjecture?

Q16 Can you find more than one set of slider positions that correspond to the transformation in any of these challenges? If so, what specific properties of the function allow you to match it by multiple sets of slider positions?

Q17 What other classes of functions have more than one set of slider positions corresponding to a single specific transformed image?

Q18 If the scale factors S_x and S_y represent the coordinates of a point S, how will the image function behave as you drag point S? Create a sketch in which you can test your conjecture.

Reflection of Functions

When you look in the mirror, you see your reflection—a mirror image of yourself. It tells you what you want to know about how you look. What if you look at a function in a mirror—what will this tell you? In this activity you will reflect the graph of a function across a line in various ways.

INVESTIGATE REFLECTIONS

The diagram at right shows pre-image function $y = f(x)$ and two reflected image functions, $g(x)$ and $h(x)$.

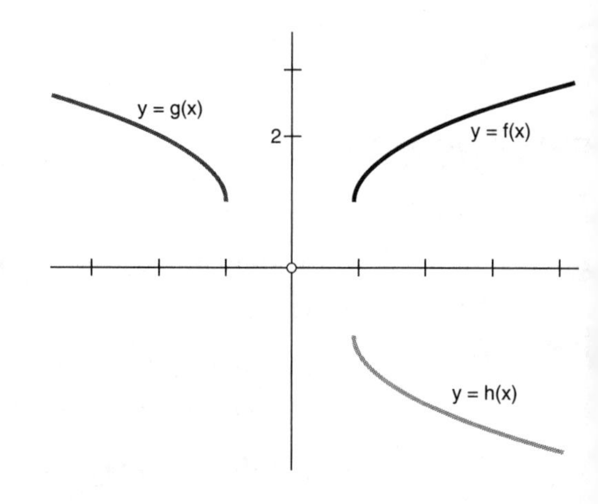

Q1 Describe the changes you see in the graphs of $y = g(x)$ and $y = h(x)$ compared to the pre-image function plot $y = f(x)$.

1. Open **Reflection.gsp** from the **1 Function Transformations** folder. Page 1 contains point P in the first quadrant.

To reflect P across the y-axis, first select the axis and choose **Transform | Mark Mirror.** Then select P and choose **Transform | Reflect.**

2. Reflect P across the y-axis, and label the image point Q. Then reflect P across the x-axis, and label the new image point R.

3. Turn on tracing for points P, Q, and R by selecting them and choosing **Display | Trace Points.** Drag point P around, and observe the traces left by all three points.

Q2 How do traces left by points Q and R compare to the trace left by the point P? Record your observations.

4. Plot the function $f(x) = 1 + \sqrt{x - 1}$. Make the plot thick, and color it with a bright color.

With P and the plot selected, choose **Edit | Merge Point to Function Plot.**

5. Merge point P to the function plot.

6. Erase the traces by choosing **Display | Erase Traces.** Animate P and observe the traces left by points Q and R while P moves along the function plot.

Q3 Can you think of a mathematical relationship between the coordinates of points P and Q? P and R?

To measure coordinates, select the point and choose **Measure | Coordinates.**

7. To explore this question, measure the coordinates of each point. Tabulate the three coordinate measurements by selecting them and choosing **Graph | Tabulate.**

8. Animate point *P*. While *P* is moving, double-click the table to record the coordinates. Collect six rows of data in the table, and compare the coordinates of the three points.

Q4 Can you think of an equation that describes the function traced by point *Q*? By point *R*? Record your equations, and test them by plotting functions.

9. To verify your equations, go to page 2 of the sketch. This page contains a coordinate system and points *P*, *Q*, and *R*. Follow the instructions on the page, and plot three new functions, $g(x)$, $h(x)$, and $v(x)$.

Q5 Which function corresponds to a reflection across the *x*-axis? Across the *y*-axis? Explain why by making a connection between algebraic operations and geometric operations.

Q6 What geometric transformation would you need to perform on point *P* in order for the image point to trace the plot of the third function?

10. To test your answer to Q6, construct a geometric image of *P* according to your prediction. Label this image *S*. Turn on tracing for point *S*, and observe its trace while moving point *P* along the function plot.

Q7 What if you changed the expression for $f(x)$? Will the relationship between geometric and algebraic transformations remain true?

To edit a function, double-click the function and enter a new expression in the Edit Function Calculator.

11. Edit the expression for $f(x)$. Check the generality of your result for several different functions, recording each time what function you tested and your observations.

EXPLORE MORE

Follow the instructions on the Challenge pages of the sketch to explore the answers to the following questions.

Q8 What combination of transformations will reflect a function across a horizontal or vertical line that is not a coordinate axis?

Q9 Is there a relationship between dilation and reflection?

Absolute Value of Functions

$|a| = a$ if $a \geq 0$, and $|a| = -a$ if $a < 0$.

The absolute value operation maps the set of real numbers to the set of non-negative real numbers. In this activity you will explore the effect of the absolute value operation in functions and its relationship to reflection.

EXPLORE ABSOLUTE VALUES

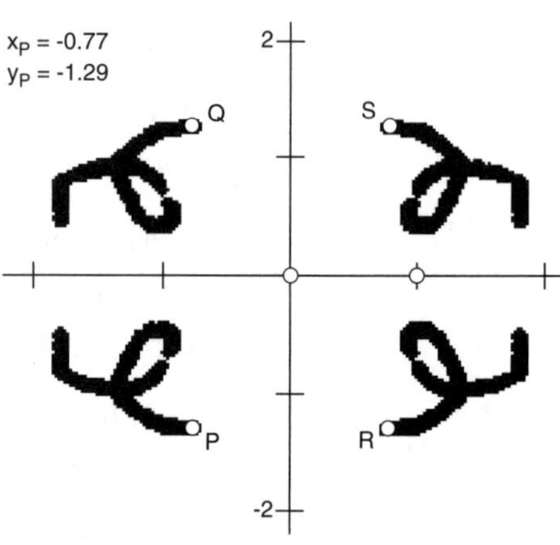

1. Open the file **Absolute Value.gsp** from the **1 Transforming Functions** folder. You will see a coordinate system and point *P*.
2. Reflect *P* across the *x*-axis, and label the reflected image *Q*. Similarly, reflect across the *y*-axis, and label the image *R*. Finally, reflect *P* across both axes, and label the image *S*.
3. Turn on tracing for all four points, and drag *P*. Observe the behavior of the reflected images.

To turn on tracing, select the points to trace and choose **Display | Trace Points.**

4. Erase the traces. Plot $f(x) = -x^2 + x + 2$, and merge *P* to the function plot. Drag *P* to observe the behavior of the reflected images.
5. Turn off tracing, and erase the traces. Plot the function $g(x) = |f(x)|$, and make the plot thick.

To turn off tracing, select the points being traced and choose **Display | Trace Points.**

Q1 Drag *P* along *f*, and observe the positions of all four points. Although it takes only one point (*P*) to trace function *f*, notice that it takes two points to trace function *g*. Which points are they? Put the labels of these two points in the *g* row on a copy of the following table.

point(s)	trace(s) function	when	and point(s)	trace(s) function	when
P	*f*	$x_P \geq 0$	*P*	*f*	$x_P < 0$
	g			*g*	
	h			*h*	
	j			*j*	

Q2 In terms of the values of x_P or y_P, determine when each of these two points is tracing function *g*. Put your answers in the table, in the columns labeled "when."

The "when" entries in the row for function "*f*" are arbitrary. They are filled in to provide a pattern for you to follow in other rows of the table.

Q3 Explain in terms of reflection and function transformation why these two points trace *g* and why each point follows the function when it does.

6. Hide the plot of g. Plot function $h(x) = f(|x|)$, and make it thick.

Q4 Drag P again, and observe which two points trace function h. Fill in the appropriate row in the table, and explain why these two points travel along h and why each follows h when it does.

Q5 Hide the plot of h. Plot $j(x) = |f(|x|)|$, and make it thick. Fill in the row of the table for function j, and explain why and when specific points trace j.

Q6 Are your explanations correct for a different function? To test your answer, enter a new expression for $f(x)$, show the plots of all the functions, and record your observations.

EXPLORE MORE

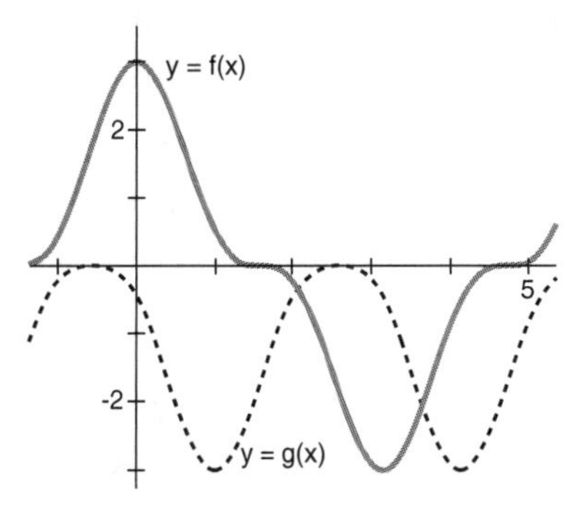

On each Challenge page of the sketch are two functions: $f(x)$ (thick and blue) is the pre-image function, and $g(x)$ (dashed) is an image function. The transformation from f to g may involve translation, scaling, reflection, and absolute value operations. Your challenge is to figure out a sequence of these transformations that will transform f into g.

Each challenge involves a number of transformations, so you may want to use one transformation first (for example, dilation) and then apply the next transformation to the result. In this way you can use function composition to get closer to the goal with each new transformed function.

You may find it convenient to use the parameters x_V and y_V to specify a translation and S_x and S_y to specify a dilation. Then you can easily drag point V or point S to adjust your translation or dilation.

To use these tools, press the **Custom** tool icon at the bottom of the Toolbox, choose the tool, and then click on the algebraic form of the function you want to transform.

For your convenience, the sketch also contains four custom tools you can use to apply reflection and absolute value transformations.

Composition of Functions

When the output of one function is used as an input for another function, the resulting output defines a *composite function.*

The increasing carbon dioxide concentration in Earth's atmosphere is causing global air temperatures to rise. Call this function *f*. The rising temperature (global warming) in turn is causing Earth's ice caps to melt. Call this function *g*. Mathematically, the melting rate of the ice caps is a *composite function* of the carbon dioxide concentration.

You can write this composite function as either $y = g(f(x))$ or $y = (g \circ f)(x)$. Function $f(x)$ is called the inside function and function $g(x)$ is called the outside function.

In this activity you will explore properties of composite functions.

EXPLORE COMPOSITE FUNCTIONS

A *dynagraph* lets you drag the input value of a function on the upper axis and shows the output value on the lower axis.

1. Open **Composition.gsp** from the **1 Function Transformations** folder. The upper dynagraph on page 1 represents *f*, the function describing how temperature depends on CO_2 concentration. The lower dynagraph represents *g*, the function describing how the melting of the ice caps depends on temperature.

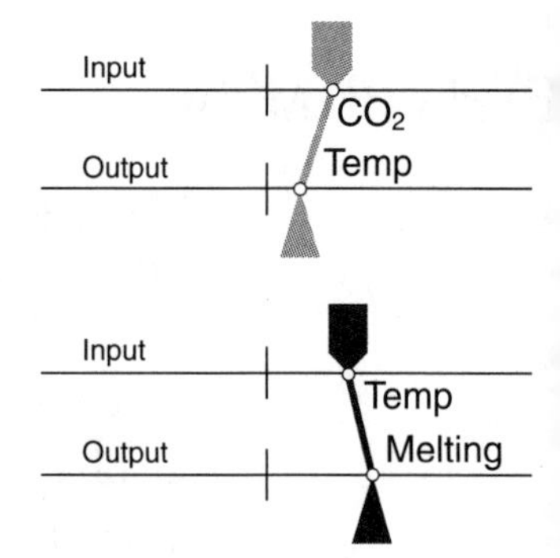

2. Drag the point CO_2 on the input axis for the first dynagraph, and observe how CO_2 concentration affects air temperature. Then drag the point on the input axis for the second dynagraph, and observe how temperature affects melting rate.

Q1 To observe the effect of CO_2 on the melting of the ice caps, how must the input of the second dynagraph relate to the output of the first dynagraph?

3. Press the buttons in the sketch to show the two hints. Follow the instructions to connect the input of the second dynagraph to the output of the first.

4. On page 2 are dynagraphs for $f(x)$, $g(x)$, and $h(x)$. Drag the input value for each function and observe the function's behavior.

5. To investigate the composition of $f(x)$ and $g(x)$, you must attach the input of *g* to the output of *f*. Press *Show Construction 1* to transfer the output value for *f* to the input axis for *g*. Point *P* is the desired input point for *g*.

6. To attach the input for *g* to point *P*, first select *B* and choose **Edit | Split Point From Axis.** Then select both *B* and *P* and choose **Edit | Merge Points.**

7. Drag the input for *f*, and observe the behavior of the linked dynagraphs.

Q2 Do these linked dynagraphs show $f(g(x))$ or $g(f(x))$? Justify your answer.

To transfer the output of g to the input axis of h, press *Show Construction 2.*

8. Similarly, attach the input of h to the output of g.

Q3 Drag the input for f, and observe the behavior of the composite function. Record your observations. Does the output value always exist? If not, what is the domain for which it does exist?

Q4 Express symbolically the function represented by the linked dynagraphs. Write your answer in both $f(g(x))$ and $(f \circ g)(x)$ forms.

To edit a function, double-click the **Arrow** tool on it.

Q5 Edit the definitions of $f(x)$, $g(x)$, and $h(x)$ to explore the composition of three different functions. Record the functions that you tried and your observations.

DOMAIN, RANGE, AND COMMUTATIVITY OF COMPOSITE FUNCTIONS

Q6 On page 3 are dynagraphs of two functions with restricted domains. Drag the input pointers of both functions. What do you think are the domain and range of the composite function $(g \circ f)(x)$?

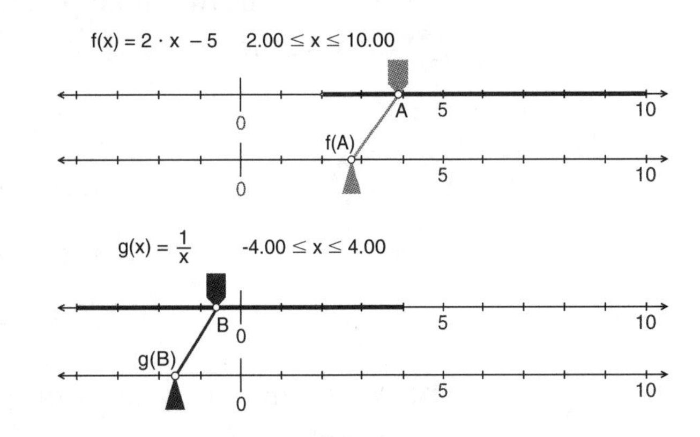

To use the **Undo All** command, press the Shift key before pulling down the Edit menu.

9. Choose **Edit | Undo All** to return the input pointers to their original positions. To create the composite function, press the *Show Construction* button to transfer the output value for f to the input axis for g, and then split B from its axis and merge it to intersection C.

Q7 Drag the input pointer for f, and record the actual domain and range of the composite function.

Q8 To investigate the composite function $g \circ f$ for each pair of functions f and g below, edit the function definitions and drag the domain end points. Determine the domain and range of each composite function.

$$f(x) = x - 3 \text{ for } 2 \le x \le 7 \quad \text{and} \quad g(x) = -2x + 8 \text{ for } 1 \le x \le 5$$

$$f(x) = 5 - x \text{ for } 0 \le x \le 7 \quad \text{and} \quad g(x) = -x^2 + 8x - 4 \text{ for } 1 \le x \le 6$$

Q9 Is composition of functions commutative? For example, does $f(g(x))$ behave the same way as $g(f(x))$? Record your conjecture before testing it in the following steps.

10. On page 4 of the sketch, the top group of axes shows the composite function $g(f(x))$, and the bottom group shows $j(h(x))$. Edit the function definitions for h and j so that $j(x) = f(x)$ and $h(x) = g(x)$.

Q10 Drag the input for each of the composite functions. Are the two composite functions equivalent? Is function composition commutative? Justify your answer in terms of your observations.

EXPLORE MORE

Q11 Is it possible to find two functions $f(x)$ and $g(x)$ such that $f \circ g = g \circ f$? To investigate, edit the function definitions on page 4 to match the functions below. Then set the two input pointers to have the same value, drag both inputs, and observe the results.

$$f(x) = 0.5x \quad \text{and} \quad g(x) = x - 2$$

$$f(x) = x^2 \quad \text{and} \quad g(x) = x^3$$

$$f(x) = x^2 \quad \text{and} \quad g(x) = \sqrt{x}$$

$$f(x) = 2x + 1 \quad \text{and} \quad g(x) = \frac{x - 1}{2}$$

Q12 What can you say about the properties of the functions f and g when $f(g(x)) = g(f(x))$?

Now you will explore graphs of composite functions on the coordinate axes and compare your observations with your previous results.

Choose **Graph | Plot New Function,** and click on the functions in the sketch.

11. On page 5, plot the same two functions you investigated on page 3: $f(x) = 2x - 5$ and $g(x) = \frac{1}{x}$. Also plot the composite functions $u(x) = f(g(x))$ and $v(x) = g(f(x))$. Make the plots of u and v thick, and color them with bright colors.

Q13 Compare plots of $u(x)$ and $v(x)$. How do your observations relate to your conclusions from analyzing dynagraphs?

Q14 Page 6 uses Sketchpad's Locus construction to construct geometrically the graphs of two functions f and g. Follow the strategy suggested on this page to construct the graph of the composite function $g(f(x))$.

Inverses of Functions

A function defines a mathematical relationship that allows you to find an output for a given input. In many real-life applications we are faced with the opposite task: we need to restore the input from the output. Doctors look at the symptoms to determine the illness. Archaeologists research fossils to learn about life on Earth a long time ago. By studying the radiation from distant stars, physicists learn about our galaxy billions of years ago. Forensic scientists find criminals from the evidence left at the crime scene. If you can restore the input from the output of the original function *uniquely*, the mathematical relation is called the *inverse* of the original function.

In this activity you will determine when inverse functions exist and explore their properties.

EXPLORE INVERSE RELATIONS

The inverse function g is often written as $f^{-1}(x)$.

If a function f has an inverse g, then $g(f(x)) = f(g(x)) = x$ for all x in the domain of f. You will use your knowledge of function composition to explore inverse functions.

To show the composition $g(f(x))$, the dynagraphs are constructed so that the input of g is always equal to the output of f.

1. Open **Inverse.gsp** from the **1 Function Transformations** folder. The first page contains dynagraphs showing the composition $g(f(x))$.

Q1 Drag x along the upper axis, and compare the values of x and $g(f(x))$. Is $g(f(x))$ ever equal to x? If so, when? Is g the inverse of f?

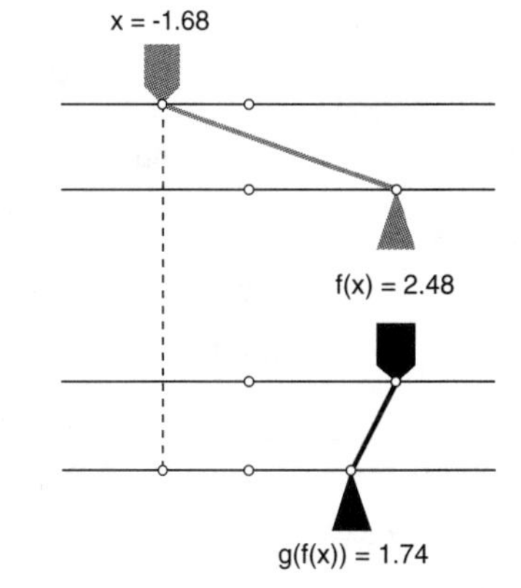

Q2 Press the *Case 2* button to change the functions. Drag x again. When (if ever) are x and $g(f(x))$ equal? Is g the inverse of f?

Q3 Check the remaining cases to determine when (if ever) x and $g(f(x))$ are equal and whether g is the inverse of f.

2. The second page of the document shows linked dynagraphs for $f(x) = 0.5x$ and $g(x) = -2x + 1$. Drag pointer x along its axis.

Q4 Is function g an inverse of function f? If not, can you think of a function g that is an inverse of f?

3. Edit $g(x)$ according to your answer to Q4, and compare the values of x and $g(f(x))$ now. Record the expressions for the two functions f and g.

To edit a function, double-click it with the **Arrow** tool.

4. Edit $f(x)$ to be the first function from the following list. Then try to edit $g(x)$ so that it's the inverse function for $f(x)$. Test your result by dragging x, and revise $g(x)$

if necessary. For each function f, record the expression for g that is the inverse.

$$f(x) = 2x - 1$$

$$f(x) = x^2$$

$$f(x) = x^3$$

Recall that an inverse function exists only when you can restore the input of the original function from its output *uniquely*.

Q5 Were you able to find an inverse function for each of the three expressions of f? If not, explain why not.

Q6 Compare the domain and range of f and g for each case in which you were able to find an inverse function.

Q7 Based on your observations, can you think of an algebraic method to determine the inverse of a function from a given function? In other words, can you find an expression for g if you know the expression for f?

EXPLORE PLOTS OF INVERSE FUNCTIONS GEOMETRICALLY

Select point P, and choose **Measure | Abscissa (x).** Then select P again and choose **Measure | Ordinate (y).**

5. On page 3 of the document, plot the function $f(x) = 0.5x - 1$. Construct point P on the function plot, and measure its x- and y-coordinates. The value of x_P is an input for function f, and y_P is an output.

Q8 Assume the inverse function g exists. If you use y_P as input to g, what will be the output? (To answer this question, consider the definition of inverse functions.) Express your answer as an ordered pair.

By plotting this ordered pair, you will plot a point on the inverse function g.

Select points P and R, and choose **Construct | Locus.**

6. Plot the point (y_P, x_P), and label it R. Drag point P, and observe the motion of point R. Point R traces out the inverse function g.

7. Construct the locus of R as P moves along the graph of f. Make it dashed.

Q9 Can you figure out an equation for the locus of point R? Explain your choice.

8. Plot the function you chose in Q9 to verify your equation. Label this function g. Make the plots of f and g thick, and color them with bright colors.

Q10 Does the plot of g match the locus of R? If so, $g(x)$ is an inverse of $f(x)$. If not, edit function g and try again.

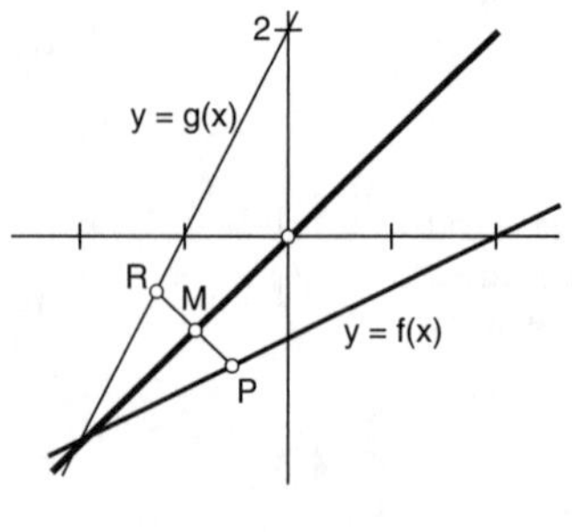

9. Construct segment PR. Drag P, and observe the slope of segment PR.

Q11 What is the slope of this segment? How can you explain this observation?

10. Construct the midpoint M of segment PR, and turn on tracing for it. Drag P, and observe the trace.

Q12 What is the equation for the trace of point M? Why does this make sense?

11. Construct the composite function $f(g(x))$, and compare the plot of the composition with the trace of point M. Record your observations.

Q13 What is the geometric relationship between segment PR and the locus of M?

Q14 Looking at the plots of f and g and the locus of M, can you think of a geometric transformation of the plot of f that will result in the plot of g? What is the role of the locus of M in this transformation?

Q15 Check the generality of your answer by trying several of these functions for f. Each time you edit f, you must also edit g to be the inverse of f.

$f(x) = x + 1$ $\quad$ $f(x) = -2x - 2$ $\quad$ $f(x) = -x + 2$ $\quad$ $f(x) = x^2$

CONCLUSION

If the inverse of a function f is also a function, then f is called *invertible.* In this case you can write the inverse function as $g = f^{-1}$. You can use the vertical line test to determine whether a mathematical equation describes a function.

Q16 Can you think of a test for the plot of the function f to determine whether the inverse will be a function as well?

Press the *Show Hint* button if you need help.

12. To explore this question, go to page 4 of the document. This page contains a plot of $f(x) = x^2$ and the identity line. Construct a plot of the inverse of f using the geometric transformation you discovered in this activity.

13. Construct a vertical line through a point on the plot of the inverse, and determine whether the inverse of f is a function.

Q17 What would be the pre-image of this vertical line for the original function?

14. Construct this pre-image and test the original function. Check the generality of your answer by trying several different expressions for function f.

EXPLORE MORE

Can you think of a function that is its own inverse? Explore this question using Sketchpad, and record your observations. Try finding your own functions. The Explore pages allow you to investigate families of linear functions and families of rational functions.

Transformation Challenge

You have learned many ways of transforming functions. You have also learned how the algebraic formulation of these transformations corresponds to the geometric definitions of the same transformations.

In this activity you'll apply your knowledge to deduce the algebraic expression for a transformed function from the graphs of the pre-image and image functions.

THE GAME

Work in groups of two or three.

Open the sketch **Transformation Challenge.gsp.** This sketch shows the algebraic expression and plot of a pre-image function $f(x)$ and shows the plot only of a transformed image $g(x)$. The sketch also shows the algebraic expression and plot of a trial function $h(x)$.

Use the *Show Expression for g* button to show and hide the formula for $g(x)$.

Select one student to go first. This student secretly shows the algebraic representation of $g(x)$, modifies it, and then hides it again.

The other participants work together to edit the trial function $h(x)$ so that its plot duplicates the plot of $g(x)$.

Once $h(x)$ precisely matches $g(x)$, the turn is over, and another member of the group takes her or his turn to create a new problem.

At each round, participants must agree on what transformations are allowed for the current round.

As play continues, the problems should get harder. Here are some possible levels of play:

- Translation only
- Dilation only (horizontal, vertical, or both)
- Reflection across an axis
- Reflection across an arbitrary horizontal or vertical line
- Absolute value operations
- Translation and dilation combined
- Dilation and reflection
- Translation, dilation, and reflection
- Translation, dilation, reflection, and absolute value

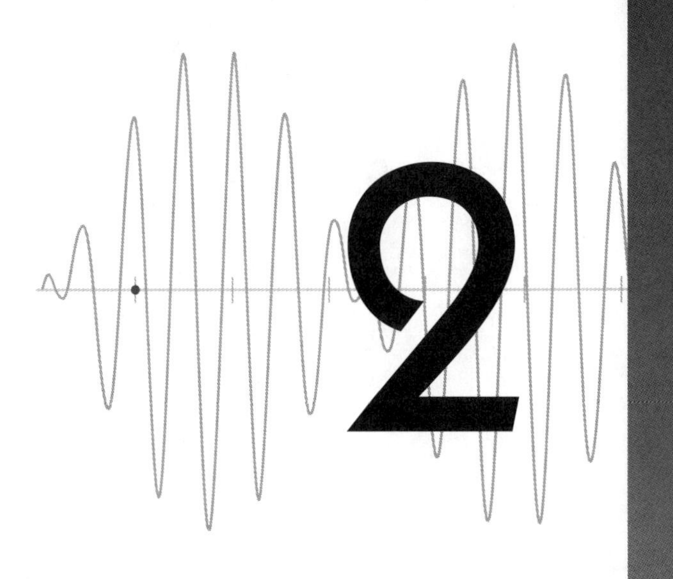

Circular Functions

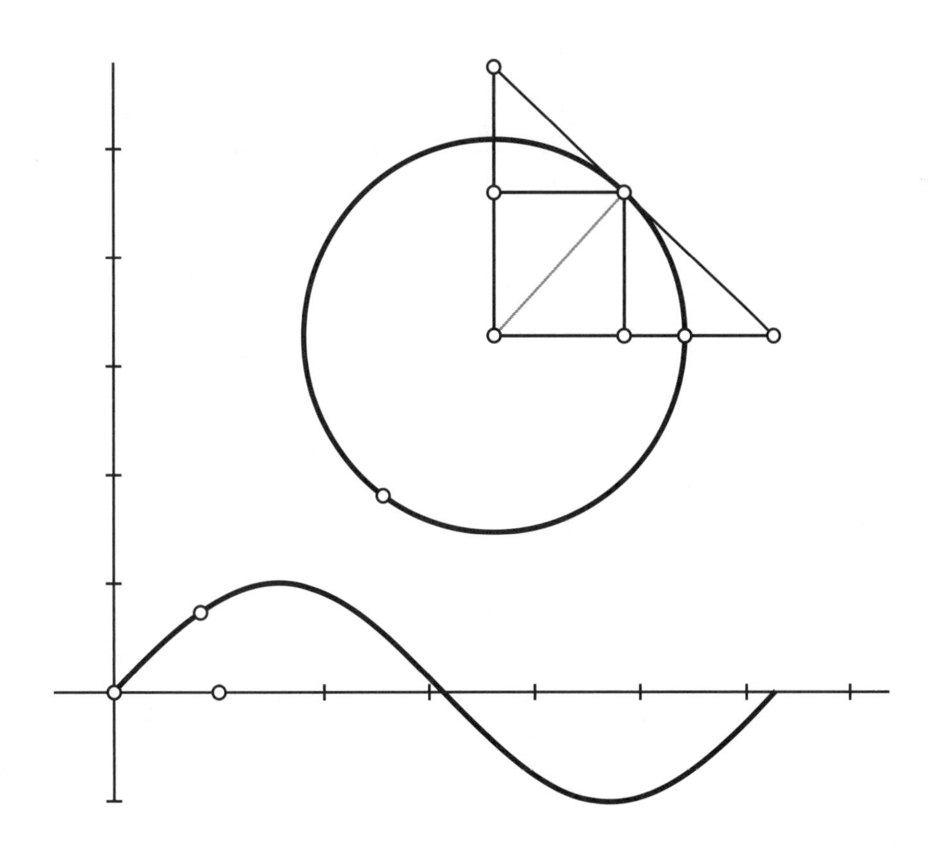

Introduction to Radians

In this activity, you'll use an eye-catching Sketchpad animation to learn about radians and discover an interesting connection to estimating the value of π.

SKETCH AND INVESTIGATE

1. Open page 1 of **Radians.gsp** from the **2 Circular Functions** folder.

2. Press the *Animate Points* button. Point A will travel along the circle's radius, and point B will move along the circle's circumference.

When point A reaches the circle's circumference, press the button again to stop the animation. If your timing is off, you can press the *Reset* button and try again.

To measure the arc length, select **Measure | Arc Length.**

3. Points A and B move at the same speed. Select the arc traced by point B, and measure its length. It should be equal, or nearly so, to the radius of the circle.

To construct a radian, you sweep out an angle whose corresponding arc length is equal to the radius of the circle. The angle is defined to be one radian.

4. Open page 2 of the sketch. Again, press the *Animate Points* button. This time, let point B travel around the entire circumference of the circle. Stop the animation when point B returns to "Start."

Notice that point A leaves a trace of its path. Each trip that point A makes from the center of the circle and back produces a petal.

Q1 How many petals are formed during point B's journey around the circumference?

Q2 Let r be the radius of the circle. For each petal formed, how far does point B travel?

Q3 Based on your answers to Q1 and Q2, how many lengths of radius r (approximately) are traced by point B as it moves once around the circumference?

Q4 Explain why your answer to Q3 makes sense based on the circumference formula, $C = 2\pi r$.

5. Press the *Reset* button to return points A and B to their original locations.

6. Start the animation again. This time, let the animation run for a while, and watch as point A traces a collection of petals. Stop the animation when point A has filled the circle with evenly spaced petals. Keep track of how many times point B travels around the circle.

Q5 How many petals did point A trace? How many times did point B travel around the circle's circumference?

Q6 Based on your answer to Q5, fill in the blanks in the following statement with integers:

__________ radii = ______________________ circumferences

Q7 Put your statement from Q6 in equation form, letting r = radius and writing circumference as $2\pi r$. Isolate π on one side of the equation.

Q8 What fraction do you obtain for π? Is this an exact value of π? If not, where might the inexactness have occurred?

Trigonometry Tracers

A periodic function is one that repeats itself over time. One way of creating a periodic function is by measuring the motion of a point that goes around a circle.

In this activity you will create a point moving around a circle and define several functions by measuring different characteristics of the point. The functions you will define are sometimes called the *circular functions* because they are generated by a point on a circle.

CONSTRUCT A MOVING POINT ON A CIRCLE

1. Start with a new sketch, and choose the **Compass** tool. Click in the middle of the sketch, and construct a circle that fills most of the screen.
2. Use the **Label** tool to label the center point *A* and the radius point *B*.
3. Select the circle, and choose **Graph | Define Unit Circle.** A coordinate system appears. Choose **Graph | Hide Grid** to hide the grid lines, leaving the axes.
4. Use the **Point** tool to construct two points: a point on the circle and the point where the circle intersects the positive *x*-axis. Label these points *C* and *D*.
5. Select the circle, point *D*, and point *C*. Choose **Construct | Arc On Circle** to construct the arc from *D* to *C*. Make the arc thick and red.

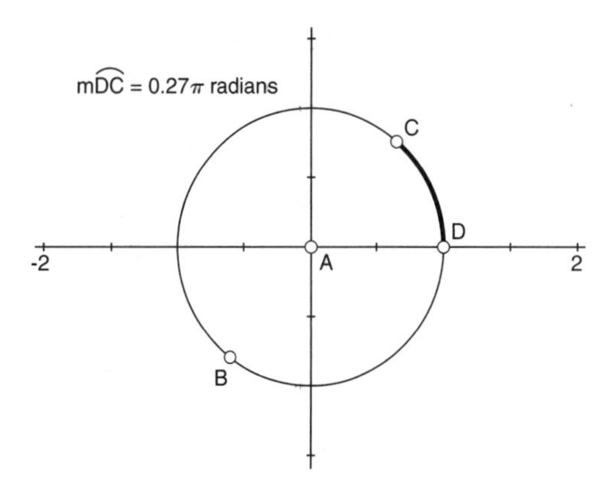

You will use the arc angle to keep track of how far around the circle point *C* is.

6. Select the arc, and choose **Measure | Arc Angle.** If the angle comes out in degrees, choose **Edit | Preferences** and change the angle units to radians.
7. Select point *C*, and choose **Edit | Action Buttons | Animation.** Accept the values in the dialog box that appears, and click OK.
8. Press the animation button that appears, to start *C* moving around the circle.

MEASURE THE POINT

By taking various measurements of the moving point, you can define certain circular functions. You will use the angle of arc *DC* as the independent variable for these functions. You will use the vertical position of *C* to define the first such function.

Press the button a second time to stop the animation.

9. Stop the animation, select point *C*, and choose **Measure | Ordinate (y).**

10. Select the two measurements in order: first the independent variable (the arc angle) and second the dependent variable (the y-value of C). Choose **Graph | Plot As (x,y).**

Q1 A new point appears. Observe its behavior as you drag point C. How is its horizontal position determined? How is its vertical position determined?

11. With this new point selected, choose **Display | Trace Plotted Point.** Press the animation button again to animate point C.

Q2 Describe the shape of the trace that results. Where does it start, and where does it end? When is the value positive, and when is it negative? (You may have to make the circle smaller, or move it to the left, to observe the entire trace. If you do change the circle, use **Display | Erase Traces** to erase the old traces so that you see only the traces that result from the new position.)

MAKE MORE MEASUREMENTS

12. Press the animation button to stop the motion, and measure the x-value (abscissa) of point C.

13. Select the arc angle measurement and the abscissa measurement in order, and plot the point determined by these measurements. Turn on tracing for the new plotted point.

Q3 Press the animation button to animate point C. Describe the behavior of the new plotted point. Where does it start, where does it stop, and how does it behave in between?

To enter a value like x_C into the Calculator, click on the measurement in the sketch.

14. Choose **Measure | Calculate,** and compute the value of y_C/x_C.

15. Select the angle measurement and the new calculated value, and plot these two values.

16. Instead of tracing, this time select C and the new plotted point in order, and choose **Construct | Locus.**

Q4 How does this new point behave? Describe the behavior in detail.

Q5 This new point shows interesting behavior as point C gets close to the y-axis. Stop the motion, and drag point C near the y-axis. Explain why the new plotted point behaves the way it does.

EXPLORE MORE

Q6 Try the following calculations: $1/y_C$, $1/x_C$, and x_C/y_C. Plot each of these values as a function of the position of C around the circle (as measured by arc angle). Describe the behavior of each of these functions.

Six Circular Functions

Trig functions are often called *circular* functions when they are defined by measurements in a unit circle.

In this activity you will create a simple diagram that contains six segments corresponding to the six circular functions. You will use these segments to calculate the values of the functions and graph them.

SKETCH AND INVESTIGATE

1. In a new sketch, use the **Compass** tool to construct a circle.
2. Use the **Label** tool to label the center point O and the radius point A.

To define the coordinate system, select the circle and choose **Graph | Define Unit Circle.**

3. Define a new coordinate system using the circle as the unit circle. Hide the grid. Construct a point on the circle anywhere in Quadrant I and label it P. Construct the intersection of the circle with the positive x-axis, and label it B.

To construct the tangent line, select P and the radius segment and choose **Construct | Perpendicular Line.**

4. Construct the radius from O to P, and construct at P a tangent to the circle. Make both the radius and the tangent dashed. Label the tangent's intersection with the x-axis Q and the tangent's intersection with the y-axis R.
5. From P construct perpendiculars to both axes. Make both perpendiculars dashed. Label the intersection with the x-axis S and the intersection with the y-axis T.

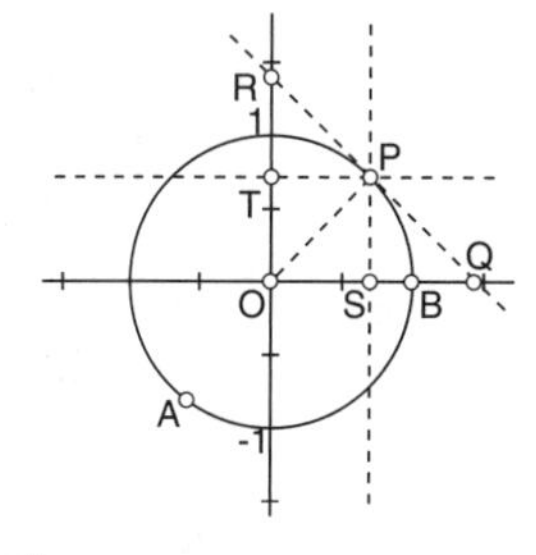

6. Hide the tangent line and perpendiculars. Construct and label the six segments listed in the following table. Make each segment thick, and give each a different color.

Segment	Label	QI	→	QII	→	QIII	→	QIV	→	QI
PS	sin	+								+
PT	cos	+	0							+
PQ	tan	+	∞	–	0	+	∞	–	0	+
PR	cot	+								+
OQ	sec	+								+
OR	csc	+	1							+

7. For each segment, select its end points and measure the Coordinate Distance.

These distances correspond to the six circular functions, but because distance is always positive, you must pay attention to the behavior of the segment to determine when the corresponding function is positive and when it is negative.

Q1 Drag point P from Quadrant I to Quadrant II, and observe each distance at the transition. Some distances are 0, some are 1, and some increase without limit. Make a copy of the preceding table, and fill in the first "→" column with 0, 1, or ∞ to indicate the behavior of the distance corresponding to each function. (Three of the cells in this column are already filled in for you.)

Q2 When a distance is 1 during the transition, the sign of the function remains the same, but when the distance is either 0 or unbounded, the sign of the function changes. In column QII, enter the new sign for each function in Quadrant II.

Q3 Observe the distances as you drag *P* from Quadrant II to Quadrant III. Fill in the next two columns of the table. Similarly, complete the rest of the table.

Signum (abbreviated sgn) is in the Calculator's Functions menu. This function returns a value based on the sign of the argument: 1 if it's positive, 0 if it's zero, or −1 if it's negative.

8. Measure the *x*- and *y*-coordinates of *P*. Use the Calculator to compute the values $\text{sgn}(x_P)$, $\text{sgn}(y_P)$, and $\text{sgn}(y_P/x_P)$.

9. Observe the behavior of these three calculations as you drag *P* into each of the four quadrants. For each distance measurement, there's one calculation that produces the desired sign for the corresponding function in your table. For instance, the calculation $\text{sgn}(x_P)$ produces the desired sign for the cos function.

10. Multiply distance *PT* by the $\text{sgn}(x_P)$ calculation, and label the result *cos.* This calculation gives the correct value of the cosine function for every position of *P.*

11. Similarly, multiply each of the other distances by the calculation that will produce the correct positive and negative values according to your table. Label each result based on the corresponding circular function.

12. Color each calculation to match the color of the corresponding segment. Hide the intermediate calculations, the measurements, and the coordinate axes.

To construct the arc, select the circle, point *B*, and point *P*, and choose **Construct | Arc On Circle.**

13. Construct the arc from *B* to *P* and measure its angle. If the measurement is in degrees, choose **Edit | Preferences** to set Angle Units to radians.

14. Construct a new point in empty space away from the unit circle, and choose **Graph | Define Origin** to define a new coordinate system. Hide the grid.

15. On the new coordinate system, plot the point $(m\widehat{BP}, \sin)$. Drag *P* to observe the behavior of the plotted point.

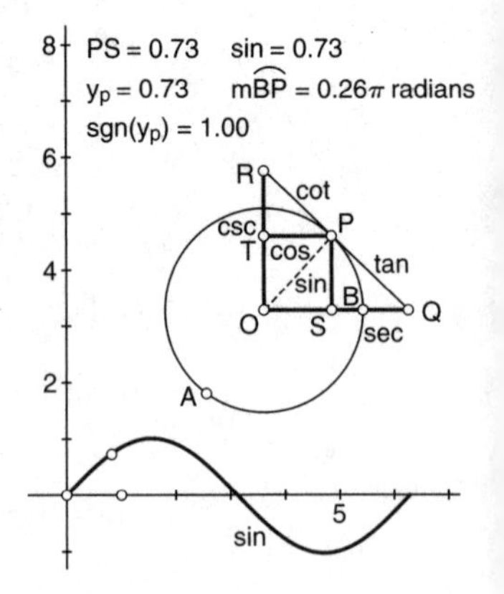

16. Construct the locus of the plotted point as *P* moves around the circle. Label this locus *sin,* and color it to match the corresponding segment and calculation.

17. Similarly, plot points and construct loci to match the other five segments.

18. For each circular function, create a hide/show button to hide or show all of its features (the segment, the calculated value, and the locus). Create an animation button to animate *P* around the circle. Use these buttons to present your work.

Transformations of Circular Functions

In this activity you will use a point on the unit circle to construct dilated images of circular functions.

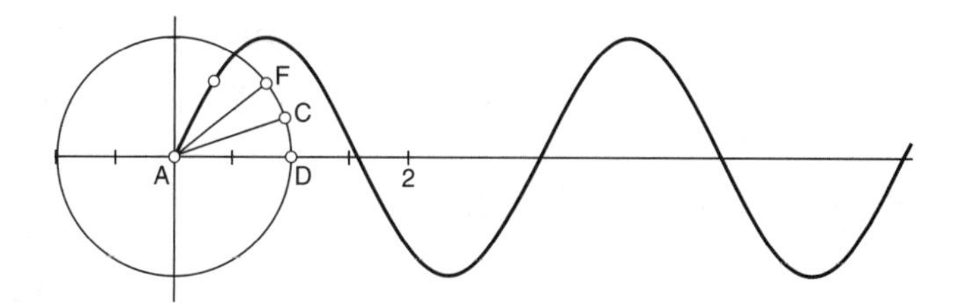

SKETCH AND INVESTIGATE

1. Open the sketch **Circular Transforms.gsp** in the **2 Circular Functions** folder. The sketch contains a parameter *k* that currently equals 2. Use the Calculator to multiply *k* by the angle measure of $\overset{\frown}{DC}$.

To mark the calculation as the angle of rotation, select it and choose **Transform | Mark Angle.**

2. Mark point *A* as the center of rotation using **Transform | Mark Center.** Similarly, mark the calculation from step 1 as the angle of rotation.
3. Rotate point *D* by selecting it and choosing **Transform | Rotate.** Label the rotated point *F*, and construct segment *AF*.

Q1 Press the *Animate Point C* button. What is the relation of $\angle DAC$ to $\angle DAF$?

Q2 For every complete trip that point *C* makes around the circle, how many times does point *F* travel around the circle?

Q3 Double-click parameter *k*, and change its value to 3. Answer Q1 and Q2 again for this new value.

4. Press the *Show Point E* button. This point, which you built in the activity Trigonometry Tracers, traces out $\sin(m\overset{\frown}{DC})$. Press the *Animate Point C* button to watch point *E* in action.

Q4 You're about to create the graph of $\sin(k \cdot m\overset{\frown}{DC})$. Before you do, make a prediction: Based on your answers to Q2 and Q3, what do you predict the graph will look like?

5. Measure y_F by selecting point *F* and choosing **Measure | Ordinate (y).**
6. Plot the point $(m\overset{\frown}{DC}, y_F)$ by selecting in order $m\overset{\frown}{DC}$ and y_F, and then choosing **Graph | Plot As (x,y).**

To turn on tracing, select the point and choose **Display | Trace Point.**

7. Label the plotted point *G*, and turn on tracing for it.

Q5 Animate *C*, and observe the trace of *G*. Is your prediction about the graph of $\sin(k \cdot m\overset{\frown}{DC})$ correct?

Choose **Display | Erase Traces** to erase existing traces.

8. Change the value of parameter *k* to draw new sine curves.

Q6 By taking new measurements, create the graphs of $\cos(k \cdot m\overset{\frown}{DC})$ and $\tan(k \cdot m\overset{\frown}{DC})$. Describe the appearance of each of these functions.

Sine Challenge

In this activity you will create a sine function whose amplitude and period are controlled by an independent point.

SKETCH AND INVESTIGATE

1. In a new sketch, choose **Edit | Preferences,** and set Angle Units to radians.

2. Construct parameters a and b. Use **Edit | Properties | Parameter** to set the keyboard adjustment for each parameter to 0.05.

3. Graph the function $y = a \cdot \sin(b \cdot x)$. Hide the grid.

4. Construct point P in a blank area of Quadrant I. Measure its x- and y-coordinates.

5. Use the + and − keys with parameters a and b to center the first crest of the sine graph at point P.

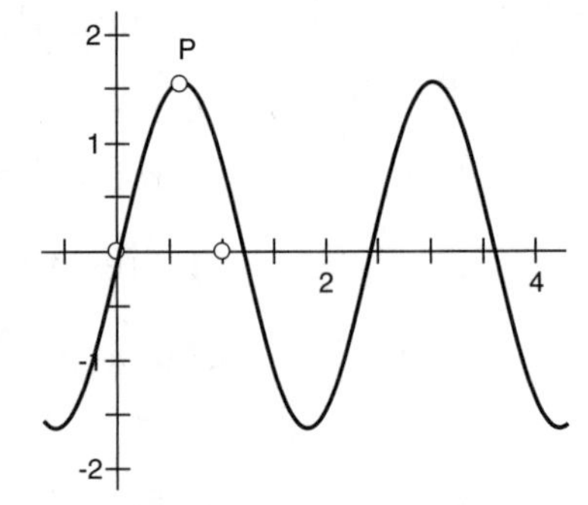

6. Select a, b, x_P, and y_P, and choose **Graph | Tabulate** to create a table. Double-click the table to make the first row permanent.

7. Move P to a new position in Quadrant I, and again adjust a and b. Double-click the table when the first crest is centered at P. Collect several more rows of data, making sure each time that the crest of the graph is centered at P.

Q1 As you look at the data in the table, what connection do you see between the value of a and the other values in the table?

Drag P all around, not just in Quadrant I.

8. Edit the function to eliminate parameter a and make the amplitude of the graph always match the position of P. Drag P to check your result.

To make a custom tool, press the custom tool icon at the bottom of the toolbox and choose **Create Custom Tool** from the menu that appears.

9. Determine how the value of b relates to other measurements, and edit the function to eliminate b while making sure that P is always at the first crest. Once you're satisfied, delete the table and parameters a and b.

10. Select point P, the function, and the graph and make a custom tool. Use the tool several times to make several easily controlled sine graphs.

EXPLORE MORE

On a new page of the sketch, use your tool to create three sine functions. Define and graph a new function that is the sum of the three existing functions. Drag the points to see how the sum of the three functions relates to the three original functions.

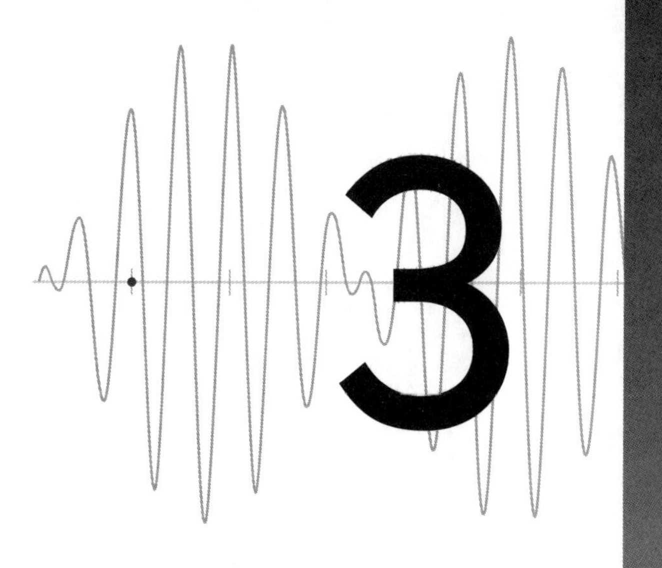

3

Trigonometric Properties

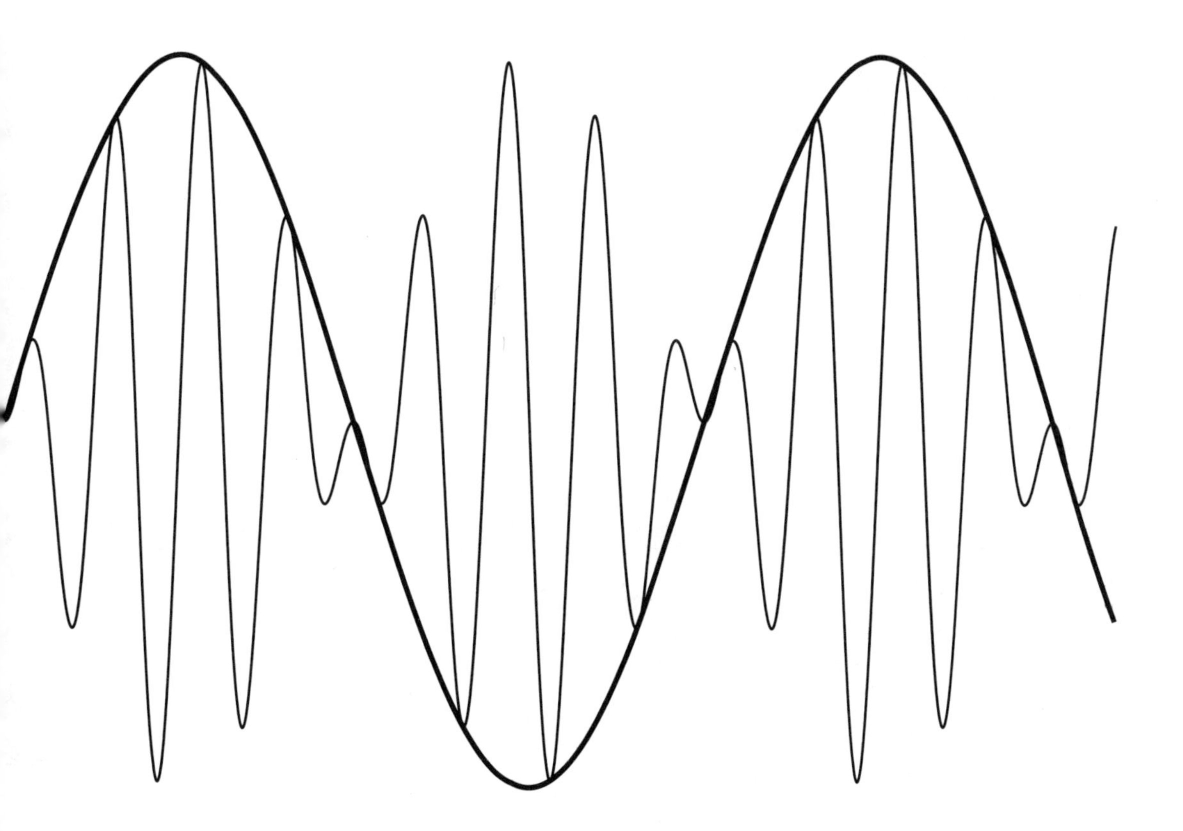

The Law of Sines

In this activity, you'll explore a triangle relationship known as the Law of Sines. Given two angles of a triangle and the length of a side, you can use the Law of Sines to compute the lengths of the remaining two sides.

SKETCH AND INVESTIGATE

Open the sketch **Law of Sines.gsp** in the **3 Trigonometric Properties** folder. You'll see two right triangles, $\triangle ACD$ and $\triangle BFE$.

Q1 The ratio of which two lengths is equal to $\sin A$?

Q2 The ratio of which two lengths is equal to $\sin B$?

1. Use Sketchpad to measure the lengths necessary to compute $\sin A$ and $\sin B$.
2. Use Sketchpad's Calculator and your measurements to compute $\sin A$ and $\sin B$.
3. Drag the two calculations next to the appropriate sine captions.

Now you'll merge segments *DC* and *EF* to turn the two triangles into one triangle.

4. Select, in order, points *F* and *C* and choose **Edit | Merge Points.**
5. Select, in order, points *E* and *D* and choose **Edit | Merge Points.**
6. Now that segment *EF* has merged with segment *DC*, notice how the labels used for the sine calculations have changed.

Q3 Solve for segment *CD* in both equations and form a statement relating $\sin A$ and $\sin B$.

Q4 Rewrite your equality from Q3 as a proportion with $\sin A$ and $\sin B$ on opposite sides of the equality.

7. Press the *Show Altitude* button to view the altitude of the triangle extending from point *A* to side *BC*.

For notation purposes, use *c* to represent side *AB* of the triangle.

Q5 Use this remaining altitude to write a new ratio representing $\sin B$ as well as a ratio representing $\sin C$.

Q6 Eliminate the common term from the two ratios in Q5. Use the resulting equality to write the complete Law of Sines statement.

You'll find the sine function under Functions on the Calculator.

Q7 Open page 2 of the sketch. Use the Law of Sines to solve this application of your result.

The Law of Cosines

In this activity, you'll explore a triangle relationship known as the Law of Cosines. Given the lengths of two sides of a triangle and the measure of the included angle, the Law of Cosines allows you to compute the length of the triangle's third side.

SKETCH AND INVESTIGATE

1. Begin with a blank sketch and draw a triangle *ABC* with angle *C* obtuse. Label the sides so that side *a* is opposite vertex *A*, side *b* is opposite *B*, and side *c* is opposite *C*.
2. On side *AB*, construct a square *ABED* that covers the triangle.
3. Construct a square *ACGF* on side *AC* of the triangle.
4. Construct parallelogram *BCGH*, based on segments *BC* and *CG*.
5. Construct square *GHEI*, based on segment *GH*.
6. Construct quadrilateral *FGID*.
7. Shade the interiors of squares *ACGF* and *GHEI* one color and shade the two parallelograms a different color.

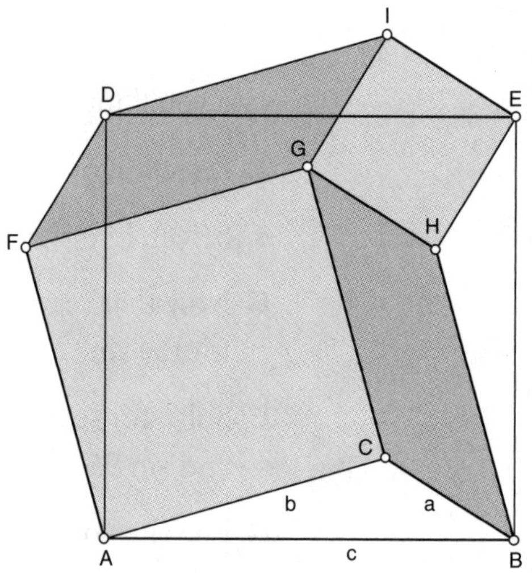

Does it surprise you to find that *E* is a vertex of the square constructed on side *GH*?

Your construction is complete and ready to be explored!

Q1 Drag point *C*, keeping angle *C* obtuse. What relationship do you notice between the two parallelograms?

Q2 Consider these two regions:

Region 1: the space occupied by the two shaded squares and the two shaded parallelograms

Region 2: the space occupied by square *ABED*

Give a convincing visual argument (no need for measurements or calculations) that the area of Region 1 is equal to the area of Region 2.

Q3 In terms of *a*, *b*, and *c*, what are the areas of the two shaded squares and square *ABED*?

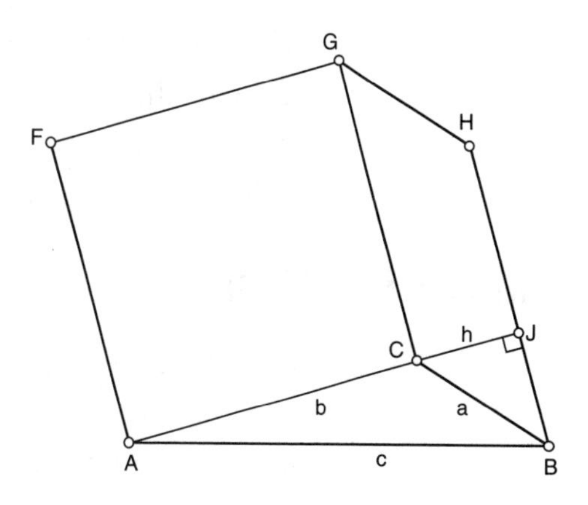

Q4 If the height of parallelogram *BCGH* is *h*, what is its area?

Q5 From Q2, you know that the area of Region 1 is equal to the area of Region 2. Use your results from Q3 and Q4 to express this relationship algebraically.

The next few questions will focus on rewriting the algebraic relationship from Q5 to include only lengths and angles from $\triangle ABC$.

Q6 If $\angle ACB = \theta$, then what is the measure of $\angle BCJ$?

member, $BC = a$.

Q7 Write an expression for $\cos(\angle BCJ)$ and then solve for h.

Q8 Use the trigonometric identity $\cos(180° - \theta) = -\cos(\theta)$ to rewrite your expression for h.

Q9 Rewrite the equality in Q5 without h. This relationship is the Law of Cosines.

Q10 In $\angle ABC$, suppose that $AC = 6$, $BC = 4$, and $m\angle ACB = 120°$. Use the Law of Cosines to find the length of $\overline{AB}$.

XPLORE MORE

Q11 In step 5, *E* turns out to be a vertex of the square constructed on side *GH*. Develop a convincing argument to explain why *E* is a vertex of this square.

Q12 In step 6, what shape does quadrilateral *FGID* turn out to be? Develop a convincing argument explaining why this must be the case.

Q13 In your Sketchpad construction, drag point *C* so that $m\angle ACB = 90°$. What happens to the two parallelograms? Explain how your picture gives a visual illustration of the Pythagorean Theorem.

Q14 Drag point *C* so that it lies anywhere along segment *AB*. What algebraic identity does your picture now illustrate?

Sums of Sinusoidal Functions

When you add two sinusoidal functions, the sum function has very interesting behavior that depends on the amplitude, period, and phase of the functions being added. You can use such functions to model many physical motions that exhibit periodic behavior. They are especially useful for analysis of waves.

SUMS OF FUNCTIONS

Begin by graphing several sine functions and examining the effects on the graph as the function parameters are varied.

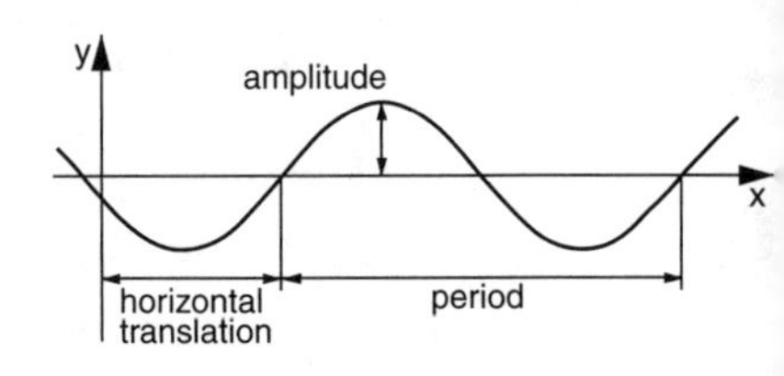

1. Open **Trig Coords.gsp** from the **3 Trigonometric Properties** folder.
2. Choose **Graph | Plot New Function** to plot the function $f(x) = \sin(x)$.

This is an elementary sine curve. The amplitude is 1, the period is 2π, and the horizontal translation is 0. (Horizontal translation is often called *phase.*)

To edit a function, double-click it with the **Arrow** tool.

Q1 What are the amplitude, period, and horizontal translation of each function below? Make a table of your predictions, and check them by editing $f(x)$.

$$f(x) = \sin(3x) \qquad f(x) = 4\sin(x)$$

$$f(x) = \sin\left(x - \frac{\pi}{2}\right) \qquad f(x) = 4\sin\left[3\left(x - \frac{\pi}{2}\right)\right]$$

Q2 Consider a sine curve in this general form: $f(x) = a \cdot \sin[b(x - c)]$. What are the amplitude, period, and horizontal translation?

The principle of *wave superposition* states that where two or more waves come together, the resulting wave is simply the sum of the components. You can observe this relationship in many natural phenomena, including fluid waves, sound, and light

Use color to match each function with its graph. Select the function definition and its graph, and choose **Display | Color** to set a new color.

Choose **Display | Line Width** to make the plot of $h(x)$ thick.

Investigate by plotting several curves on the same grid and finding their sum.

3. Edit the existing function definition to be $f(x) = 3\sin(x)$. Plot a new function, $g(x) = \sin(10x)$, and make it a different color from the previous graph. Plot another new function, $h(x) = f(x) + g(x)$. Make the plot of $h(x)$ thick, and use yet another color.

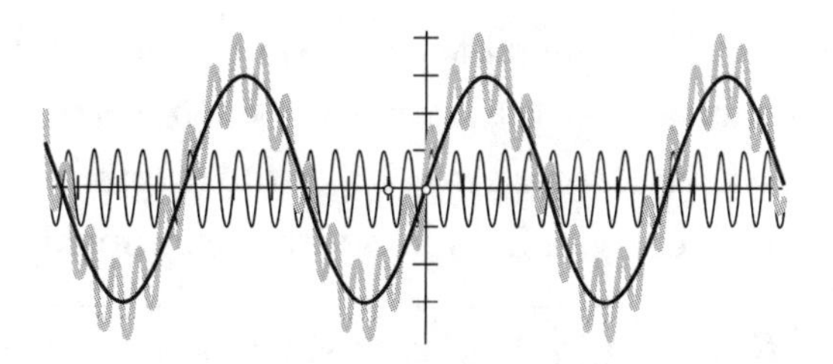

To understand the graph of $h(x)$, you can imagine that one of the sine curves acts as an axis for the other. Function h appears to be crawling along the length of f. Next you will see what happens when you combine two graphs that are more similar in appearance.

To create a hide/show button, select the objects and then choose **Edit | Action Buttons | Hide/Show.**

4. Create separate hide/show buttons for each of the three graphs. The graphs are easier to examine when you view them selectively.

5. Edit the function definitions for f and g:

$$f(x) = \sin(x) \qquad g(x) = \sin\left(x - \frac{\pi}{2}\right)$$

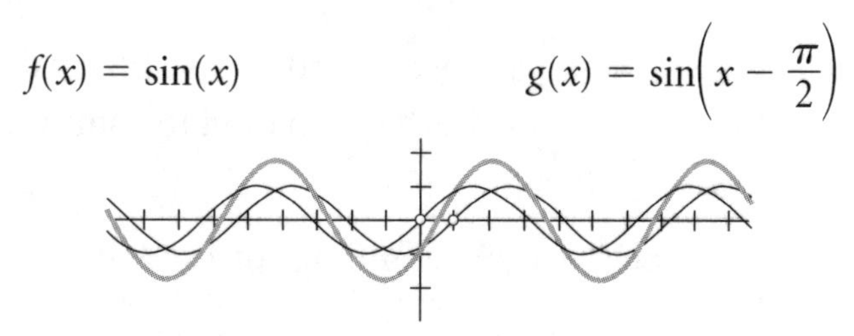

Q3 What are the period and horizontal translation for h? How do these compare with the characteristics of f and g?

6. Set these function definitions for f and g:

$$f(x) = \sin(6x) \qquad g(x) = \sin(5x)$$

This image shows only $h(x)$.

Q4 The graph of h shows clusters of waves called *wave packets,* which create an effect called *beats.* In sound waves, this is the pulsing effect that you hear when two notes have slightly different pitch. What is the length of each wave packet? The amplitude is variable. What is its greatest value?

Q5 The following table lists several different definitions for the functions in the form $f(x) = \sin(ax)$, $g(x) = \sin(bx)$. Try each combination on the sketch. In each case, record $a - b$ and the length of the wave packet in the graph of $h(x)$. What can you infer about the relationship between $a - b$ and the packet length?

$f(x) = \sin(ax)$	$g(x) = \sin(bx)$	$a - b$	packet length
$\sin(6x)$	$\sin(5x)$		
$\sin(10x)$	$\sin(9x)$		
$\sin(10x)$	$\sin(9.5x)$		
$\sin(10x)$	$\sin(8x)$		
$\sin(16x)$	$\sin(14x)$		

When you study wave behavior, the waves are usually moving. You can simulate this motion in your model.

To create a parameter, choose **Graph | New Parameter** and type the desired parameter label into the dialog box.

To create an animation button, select the object to animate and choose **Edit | Action Buttons | Animation.** Set the desired direction, speed, and domain before you click OK.

7. Create a new parameter t and set its initial value to 0. Create an animation button for t, and set the animation to continuously increase on the domain from 0 to 1000.

8. Set these function definitions for f and g:

$$f(x) = \sin[6(x - t)] \qquad g(x) = \sin[5(x - t)]$$

Q6 Press the animation button. As t increases, how do the speed and direction of h compare with the speed and direction of its components, f and g?

9. Now examine the sum of two graphs that have slightly different periods and are moving in opposite directions. Enter these function definitions:

$$f(x) = \sin[6(x - t)] \qquad g(x) = \sin[5(x + t)]$$

Q7 How do these definitions make the waves of h move in opposite directions? What changes took place in the speed and direction of travel of the wave packets? Do they have the same length?

Q8 Switch back and forth between the function definitions in steps 8 and 9, and describe and compare the wave behavior. Which definitions result in simpler behavior of h? Why does one pair of definitions result in more complex behavior?

Q9 What do you think the results will be if the two functions have the same period and amplitude, but move in opposite directions? Write down your conjecture about the amplitude, period, and motion of h.

10. Change the definitions so that the functions have the same period, but opposite directions:

$$f(x) = \sin(x - t) \qquad g(x) = \sin(x + t)$$

Q10 Animate parameter t. Was your prediction correct? Describe the motion of the graph of h. This effect is called a standing wave.

Save the sketch you created. You can use it again in the next activity.

PHYSICS CONNECTIONS

Q11 Describe something in the physical world that exhibits the standing wave behavior seen in step 10.

Q12 In step 9, the wave packets move much faster than the component parts, f and g. Some people have proposed that this effect might enable radio communications that travel faster than light. Can you identify the flaw in this reasoning?

EXPLORE MORE

Experiment by combining more than two sine functions. For instance, graph the sum of the following functions:

$$f(x) = \sin(x) \qquad g(x) = \frac{1}{3}\sin(3x) \qquad h(x) = \frac{1}{5}\sin(5x)$$

Add several more terms to this sum, following the same pattern, and describe the resulting function. If you added many more terms, what do you think the graph would look like?

Products of Sinusoidal Functions

In this activity you will explore products of sine functions and the relationship between a sum and a product of sinusoidal functions.

INVESTIGATE WAVE PACKETS

To add a duplicate copy of the existing page, choose **File | Document Options | Add Page | Duplicate | 1.** Then rename the pages to be *Sum* and *Product.*

If you need to start from scratch, begin with the **Trig Coords.gsp** sketch and plot the three functions listed in step 1.

1. Open the sketch that you created in the previous activity. You can use it again with a few alterations. Add a new page to the sketch by duplicating the existing page. Name the original page *Sum* and the new page *Product.* On the Product page, make these changes to the function definitions:

$$f(x) = \sin(x) \qquad g(x) = \sin(10x) \qquad h(x) = f(x) \cdot g(x)$$

Q1 The graph of h (above left) is again exhibiting the beat effect that you saw in the sum of two functions. When you show the graphs of f and h together (above right), notice how f outlines the wave packets of h. A function like f that encloses another plot is called an *envelope.* In what way are the graphs of g and h related?

2. Change the definitions of f and g so that they move in the same direction as parameter t changes:

$$f(x) = \sin(x - t) \qquad g(x) = \sin[10(x - t)]$$

Q2 Describe the motion of the wave packets of h as t increases. Do they have the same direction and speed as the graphs of f and g? How are the individual wave crests behaving?

3. Make this change to f so that the graphs of f and g move in opposite directions.

$$f(x) = \sin(x + t) \qquad g(x) = \sin[10(x - t)]$$

Q3 Describe the motion of the wave packets again.

4. Now go back to stationary graphs. See what happens when the functions have amplitudes other than 1.

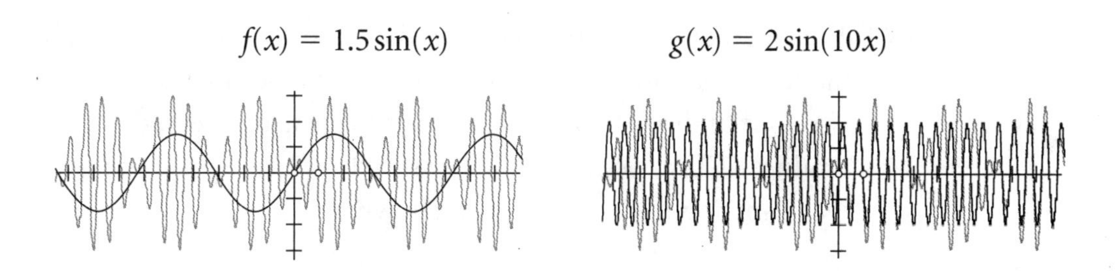

The image on the left shows the graphs of *f* and *h*. On the right are *g* and *h*. It appears that neither *f* nor *g* forms an envelope for the wave packets.

Q4 What are the amplitude and period of the envelope of wave packets? What relationship might there be between the amplitude of the envelope and the amplitudes of *f* and *g*? What relationship might there be between the periods?

NVELOPES OF WAVE PACKETS

Consider this:

$$\begin{aligned} h(x) &= f(x)g(x) \\ &= 1.5\sin(x) \cdot 2\sin(10x) \\ &= 3\sin(x) \cdot \sin(10x) \end{aligned}$$

5. From this product, take the sine function that has the greatest period—in this case, $\sin(x)$. Give it the coefficient 3, and use it to plot one more function:

$$q(x) = 3\sin(x)$$

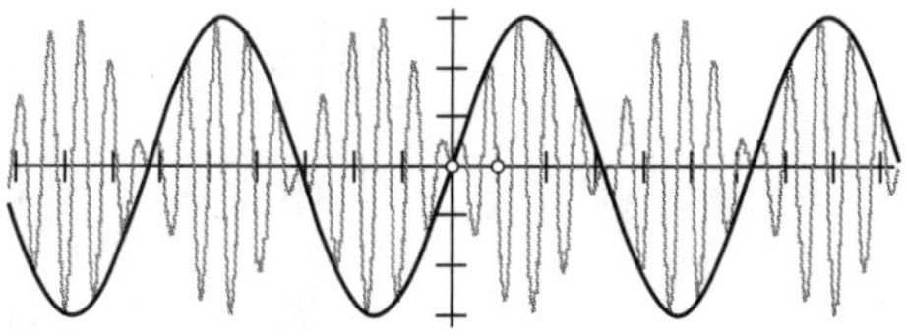

Q5 The graph of q appears to envelop the wave packets in the graph of h. If you used the coefficient 3 with $\sin(10x)$ instead, will that function also form an envelope? Use your findings to define functions that form envelopes for the following products. Check your work in the sketch by editing f, g, and q appropriately for each function $h(x) = f(x) \cdot g(x)$ shown next.

$$h(x) = 5\sin(x) \cdot 2\sin(6x)$$
$$h(x) = 3\sin(12x) \cdot 3\cos(2x)$$
$$h(x) = 0.5\cos(9x) \cdot 0.8\sin(2x)$$

You have seen the beat effect created by both the sum and the product of sinusoidal functions. To understand the connection between sums and products, consider these identities:

Note that these identities can be applied only when the component functions in the sum have the same amplitude.

$$\sin\alpha + \sin\beta = 2\sin\frac{1}{2}(\alpha + \beta) \cdot \cos\frac{1}{2}(\alpha - \beta)$$
$$\cos\alpha + \cos\beta = 2\cos\frac{1}{2}(\alpha + \beta) \cdot \cos\frac{1}{2}(\alpha - \beta)$$

You can use these identities to express the sum of sinusoidal functions as a product.

6. Back on the Sum page, define f, g, and h as follows:

$$f(x) = \sin(8x) \qquad g(x) = \sin(7x) \qquad h(x) = f(x) + g(x)$$

Use the identity to express this as a product:

$$\begin{aligned} h(x) &= \sin(8x) + \sin(7x) \\ &= 2\sin\frac{1}{2}(8x + 7x) \cdot \cos\frac{1}{2}(8x - 7x) \\ &= 2\sin\left(\frac{15}{2}x\right) \cdot \cos\left(\frac{1}{2}x\right) \end{aligned}$$

7. You now have h expressed as the product of two functions. The envelope function for this product is $2\cos(x/2)$. Define and plot another function on the Sum page to confirm that this works:

$$q(x) = 2\cos\left(\frac{1}{2}x\right)$$

Q6 Use the same identity to express the sum of the two functions $f(x) = \sin 5x$ and $g(x) = \sin 3x$ as the product of functions, and use the result to determine the envelope function. Check your results in your sketch.

SUMMARY

You can formulate the general case involving the sum of two sine functions having equal amplitude this way:

$$h(x) = k\sin(ax) + k\sin(bx)$$

Q7 Write h as the product of sinusoidal functions. What function forms an envelope for h? What is its amplitude and period? What is the length of the wave packets in h?

To confirm your answers, define appropriate functions, using your own choice of values for k, a, and b. You can easily investigate many different sinusoidal functions by creating parameters for k, a, and b instead of using specific numbers.

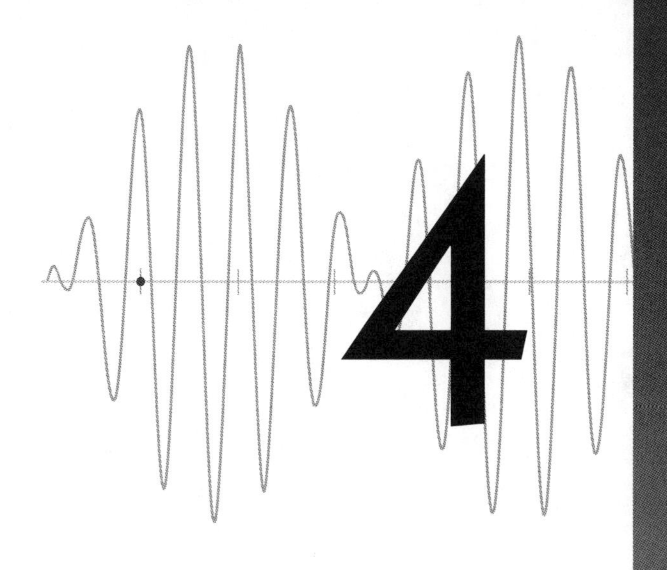

4

Other Functions

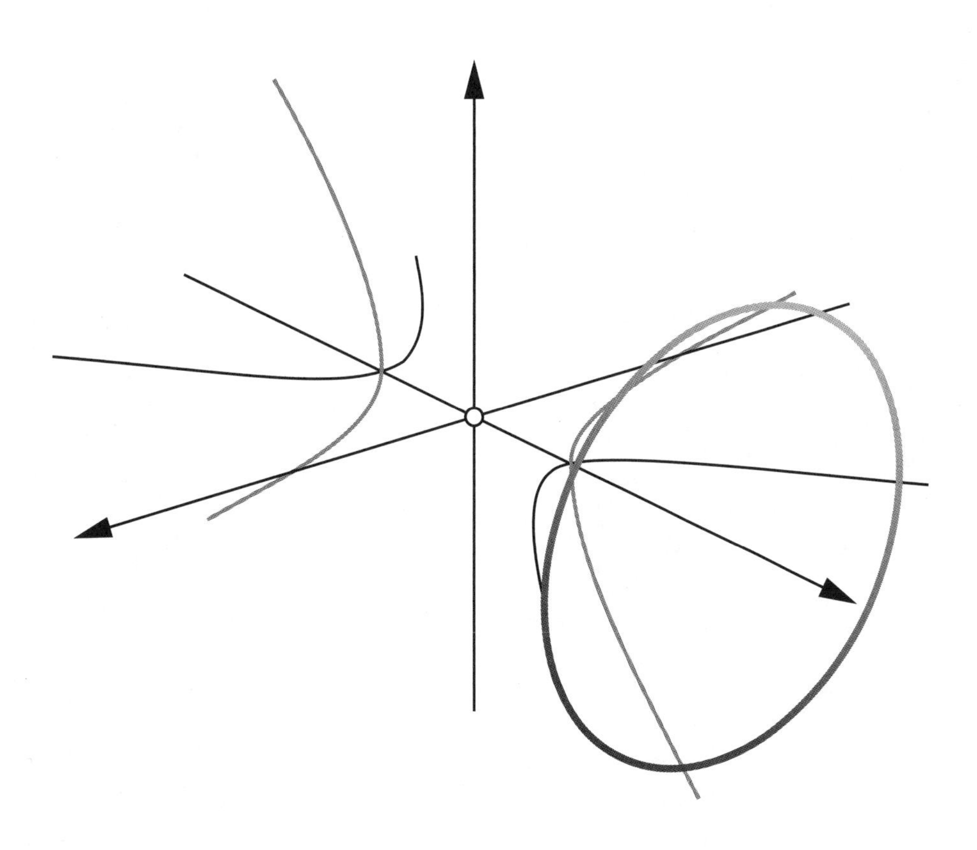

Exploring the Roots of Quadratics

This activity focuses on quadratic functions. Using a pair of linked coordinate systems, you'll explore the relationship between a quadratic's roots, its graph, and the value of its coefficients.

QUADRATIC ROOTS

1. Open the sketch **Quadratic Roots.gsp** in the **4 Other Functions** folder. You'll see two coordinate axes and the quadratic function $f(x) = (x - r_1)(x - r_2)$.

Point *R* controls the function. The *x*-coordinate of point *R* is r_1, one root of $f(x)$. The *y*-coordinate of point *R* is r_2, the other root of $f(x)$. As you drag point *R*, the roots change, which, in turn, affects $f(x)$.

Explore each question below by dragging point *R* slowly and observing the movement of the corresponding quadratic function. After you develop a conjecture, think about why such behavior is occurring.

Q1 For which locations of point *R* does $f(x)$ pass through the origin?

Q2 For which locations of point *R* does $f(x)$ intersect the *x*-axis at only one point?

Q3 For which locations of point *R* is $f(x)$ symmetric across the *y*-axis?

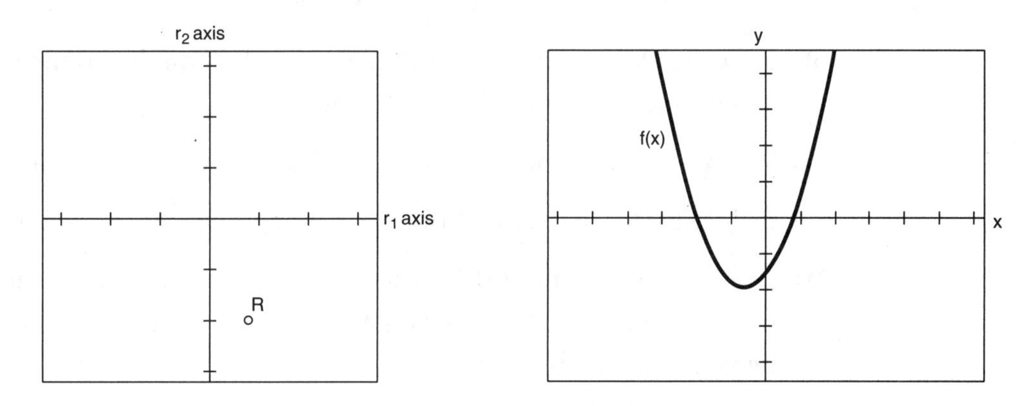

QUADRATIC COEFFICIENTS

Now you'll examine another pair of linked coordinate systems and uncover the graphical relationships between the roots of $f(x) = (x - r_1)(x - r_2)$ and the coefficients of $f(x)$ when its two terms are multiplied together.

Q4 When you expand $f(x)$, you can write it as $f(x) = x^2 + bx + c$. Complete the following equalities in terms of *b* and *c*.

$$r_1 + r_2 = \underline{\hspace{3cm}}$$

$$r_1 r_2 = \underline{\hspace{3cm}}$$

2. Open the second page of **Quadratic Roots.gsp.** You'll see point *R* on one set of axes and another point, *P*, on the other axes. The coordinates of *P* are (b, c)—the coefficients of $f(x)$.

3. For the questions that follow, it may help to restrict the movement of point *R* to a specific segment. To do so, select point *R* and the desired segment and choose **Edit | Merge Point to Segment.**

To free point *R* from its segment, select the point and choose **Edit | Split Point from Segment.**

Q5 Describe and explain the behavior of *P* as *R* travels along the r_1-axis.

Q6 Describe and explain the behavior of *P* as *R* travels along the r_2-axis.

A CONNECTION TO THE QUADRATIC FORMULA

To clear the screen, choose **Erase Traces** from the Display menu.

4. Select points *R* and *P* and choose **Display | Trace Points.** Deselect both points. Drag point *R* around the entire screen (even beyond the visible portion of the r_1-r_2 plane) and observe the trace of point *P*.

Q7 Note that the trace of point *P* does not cover the entire *b*-*c* plane. Explain why point *P* can never lie on the positive *c*-axis.

Q8 For which locations of point *R* does point *P* sit on the boundary of the traced and untraced regions of the *b*-*c* plane?

Q9 When $r_1 = r_2$, rewrite the two equalities from Q4 in terms of r_1, *b*, and *c* only.

Q10 Reduce the two equalities from Q9 to a single equality by eliminating the r_1 term.

Q11 Using your answer to Q10, write an inequality in terms of *b* and *c* describing when $f(x)$ has no real roots. Write another inequality describing when $f(x)$ has two real, distinct roots.

You can confirm and understand your answers to Q10 and Q11 in a different way by applying the quadratic formula to $f(x)$.

Q12 According to the quadratic formula, what are the roots of $f(x) = x^2 + bx + c$?

Q13 Based on your answer to Q12, when does $f(x)$ have two distinct roots? Two identical roots? No real roots?

Check to see whether your answers to these questions match the answers you gave in Q10 and Q11.

Analytic Conics

You can interpret conic sections both geometrically and analytically (using equations). Both representations are useful and interesting. Ideally you should learn both the geometric and analytic representations and never lose the connection between the two.

CIRCLE, ELLIPSE, AND HYPERBOLA IN STANDARD POSITION

A circle is simply a special case of an ellipse.

The equations for circles, ellipses, and hyperbolas are simplest when they are in standard position. In standard position, the center is at (0, 0) and the axes are horizontal and vertical. You can express the equations of these three conics in this form:

$$\frac{x^2}{A} + \frac{y^2}{B} = 1$$

To edit a parameter, use the **Arrow** tool to double-click it, and then type a new value. You can also change the value using the **+** or **–** key on the keyboard, or you can select the parameter and choose **Display | Animate.**

1. Open the Standard Form page of **Conics.gsp,** in the **4 Other Functions** folder. At the top of the screen is a standard equation of an ellipse or a hyperbola. You cannot edit the equation directly, but you can edit parameters *A* and *B* (above their corresponding values in the equation). Experiment by changing the parameters and observing the effects on the shape of the curve.

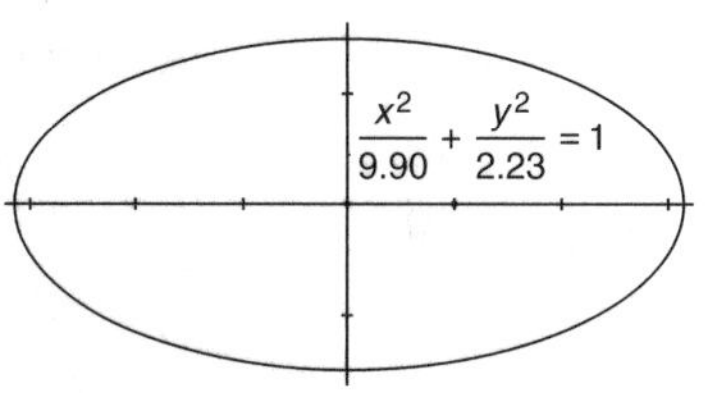

Q1 From the values of *A* and *B,* how can you determine whether the curve is an ellipse, a circle, or a hyperbola?

Q2 From the equation of an ellipse, how can you determine which is the major axis? What are the major and minor radii?

Q3 From the equation of a hyperbola, how can you determine which is the transverse axis? What are the transverse and conjugate radii?

Q4 For what values of *A* and *B* is there no solution to the equation?

Q5 Why can't you generate a parabola from an equation in this form?

GENERAL SECOND-DEGREE EQUATIONS

You can represent any conic section by a second-degree Cartesian equation:

$$Ax^2 + Bxy + Cy^2 + Dx + Ey + F = 0$$

2. Open the 2nd-Degree page of **Conics.gsp.** At the top of the screen is a general second-degree equation in *x* and *y*. You can change the equation by editing parameters *A* through *F*, which are above their corresponding values in the equation. The graph of the equation is in blue. Other objects (such as foci, center points, directrices, and asymptotes) appear when appropriate.

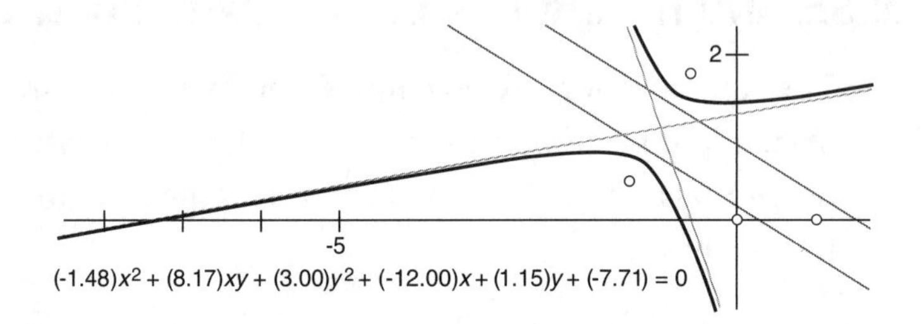

3. To begin, remove the *xy* term from the equation by setting parameter *B* to zero. Experiment by changing the other parameters and observing the effect.

Q6 What general statements can you make about the graph when *B* is zero?

Q7 The following table contains descriptions of conic sections. Either systematically or by guessing, find an equation that will produce each locus.

Locus	Equation
Circle	
Ellipse	
Hyperbola	
Parabola	
Line	
Two intersecting lines	
Two parallel lines	
Point	
No solution	

Q8 is best answered by describing the ranges of possible values for the coefficients.

Q8 Still assuming that *B* is zero, describe the general forms of the equations for the circle, ellipse, hyperbola, and parabola.

Q9 The solution is a line if you reduce the equation to the first degree by setting the first three parameters to zero. Is it possible to have a second-degree equation with a solution that is a single line? How?

4. Change parameter *B* to a nonzero number. Experiment again with changing the other parameters.

Q10 Review the generalizations that you made in Q6. Are they still true when $B \neq 0$?

Parametric Functions

In this activity you will explore the use of parametric functions to plot curves using rectangular (x, y) and polar (r, θ) coordinates. With parametric functions, rather than defining one coordinate as a function of the other, you define both coordinates as functions of a third variable, called the *parameter.*

RECTANGULAR COORDINATES

1. Open the Rectangular page of **Parametric.gsp** in the **4 Other Functions** folder. Spend a few minutes getting familiar with the objects on the screen.

On the left side of the screen are definitions for the two parametric functions. Three measurements show the current values of θ, $x(\theta)$, and $y(\theta)$. At the bottom are values you can use to set the upper and lower limits for θ. The slider controls the current value of θ. Point P represents the current position of $(x(\theta), y(\theta))$.

The parametric variable is often called t. This activity uses θ because Sketchpad's Calculator can easily create functions of θ.

Q1 Examine the function definitions. What are the coordinates of point P when $\theta = 0$? What are the coordinates when $\theta = 1$?

2. Press the *Advance* button to increase the value of θ. As point P moves, it leaves a trail behind so that you can examine the path in progress. Press the *Reset* button to return the parameter to its original value.

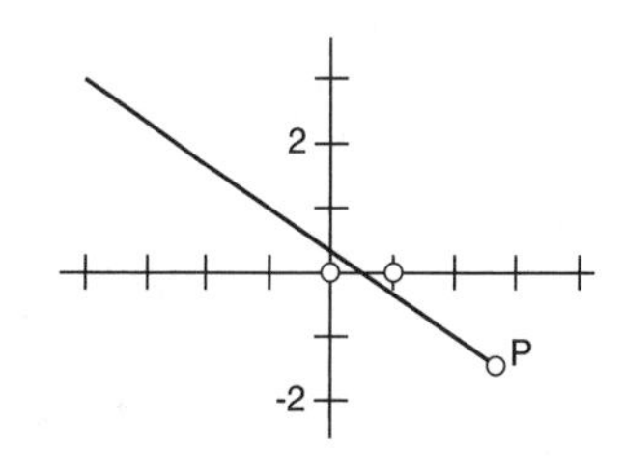

Q2 How can you edit the function definitions so that the locus is a line through points $(-5, 6)$ and $(2, 1)$? Test your answer on the sketch.

To edit a function, double-click it with the **Arrow** tool, and use the Function Calculator's keypad to change the function.

Q3 What general function definitions can you use to describe a line through points (x_1, y_1) and (x_2, y_2)?

One advantage of parametric functions is that you can use them to specify curves that you cannot specify in the form $y = f(x)$, particularly curves that conflict with the vertical line rule.

3. Press the *Lower Limit* button to reset the sketch. Then change both the functions and the limits, and plot the curve:

Double-click a limit value (like the upper limit) to change it. Type P (Windows) or option+P (Mac) to enter π.

$x(\theta) = \cos\theta$ upper limit $= 2\pi$

$y(\theta) = \sin\theta$ lower limit $= 0$

Q4 You should see a circle. What are its center point and radius? Describe the circle with a single equation of x and y.

Q5 How can you change the parametric functions so that the circle has radius 3?

4. If a unit circle is stretched horizontally by a factor of 5 and stretched vertically by a factor of 2, the new image is an ellipse with major radius 5 and minor radius 2. You can edit the functions to create this same ellipse: $x(\theta) = 5\cos\theta$ and $y(\theta) = 2\sin\theta$.

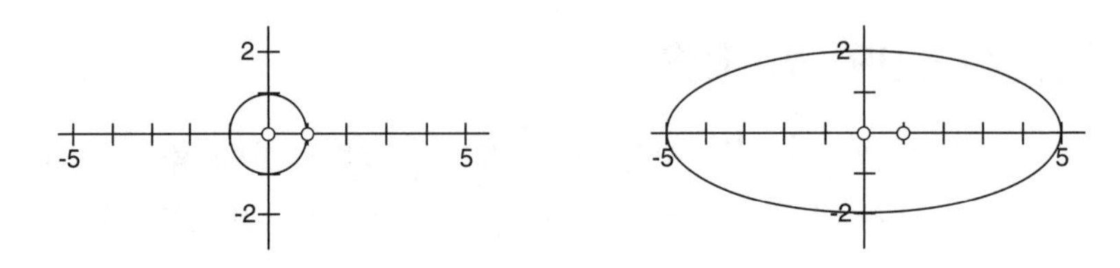

You may have seen a pendulum that traces a pattern in sand. A joint at the top of the cable restricts the movement so that the effective cable length (hence, the oscillation period) depends on whether it is swinging in the *x*-direction or the *y*-direction.

5. Enter these function definitions to simulate a pendulum swinging with different amplitudes and periods in both the *x*- and *y*-directions.

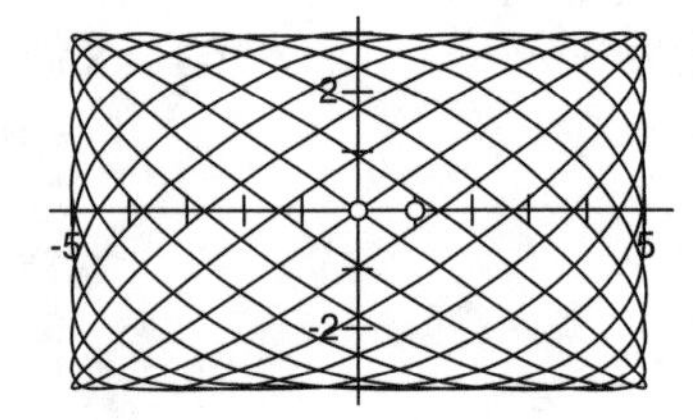

$x(\theta) = 5\sin 10\theta$ $\qquad$ $y(\theta) = 3\sin 11\theta$

Q6 What are the period and amplitude of each function?

Q7 As θ goes from 0 to 2π, how many complete cycles will $x(\theta)$ and $y(\theta)$ make?

POLAR COORDINATES

The polar sketch uses the parametric variable *x* to avoid confusion with θ.

When you define both *r* and θ as functions of *x*, you can create curves that are not possible in the forms $r = f(\theta)$ or $\theta = f(r)$.

6. Open the Polar page of **Parametric.gsp.** When you open the sketch, it should have these settings:

$r(x) = 4 + \sin(\pi/2 + 6x)$ $\qquad$ upper limit $= 2\pi$

$theta(x) = x + (\pi/8)\sin 6x$ $\qquad$ lower limit $= 0$

Q8 Draw the curve. How many rotation and reflection symmetries does it have?

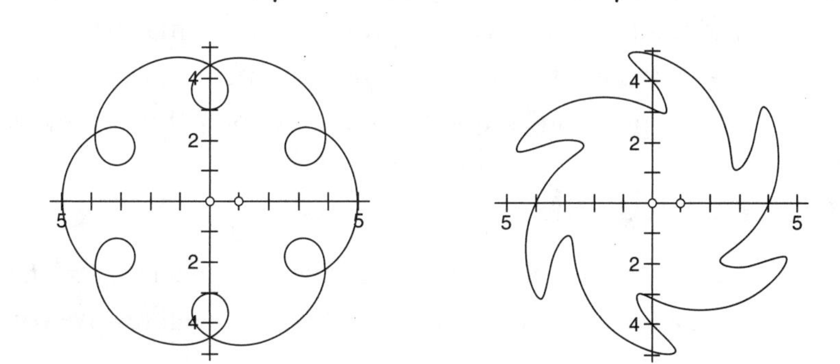

7. Make this minor change to one of the functions: $r(x) = 4 + \sin 6x$.

Q9 Again, how many rotation and reflection symmetries does the curve have?

Q10 How can you edit the functions to create a closed curve with 3 rotation symmetries but no reflection symmetries?

EXPLORE MORE

Q11–Q14 show curves that were created using **Parametric.gsp.** Attempt to duplicate them. Create new curves of your own. Identify the symmetries where they exist.

Q11

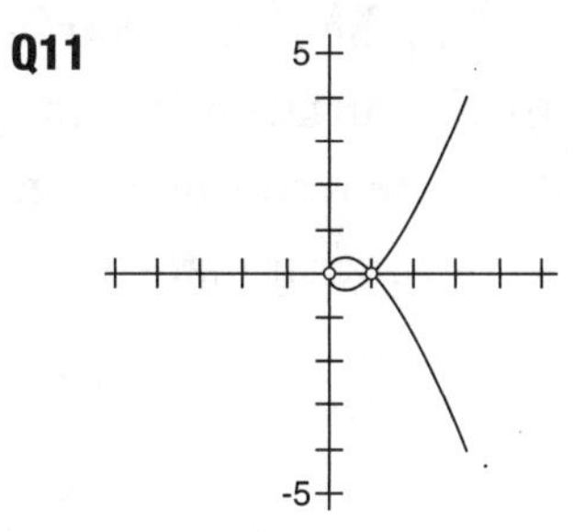

Q12

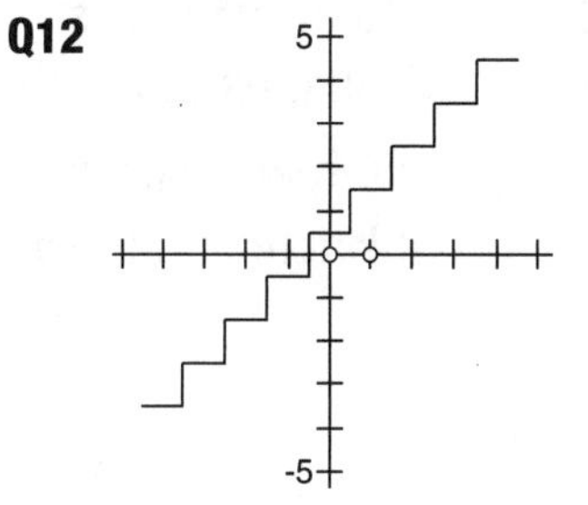

Q13

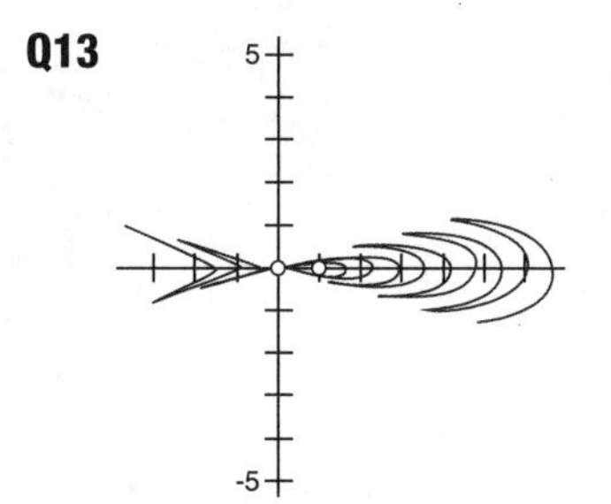

Q14

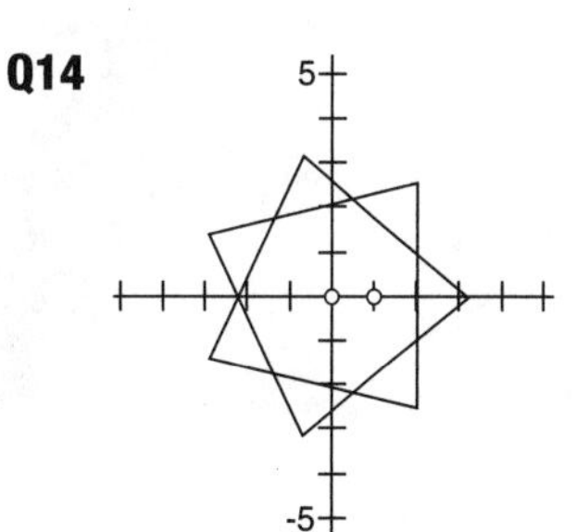

Surfaces

A *surface* is a connected set of points in three-dimensional space. It is a boundary and has no thickness or volume. A surface may be open or closed. If it's closed, it may enclose a region with finite volume. You can think of it as an extremely thin coat of paint on an object, or as extremely thin shrink-wrap on a package.

SURFACES OF REVOLUTION

You can revolve a curve about an axis to generate a *surface of revolution.* The surface is the path of the curve. In this section, you will revolve conic sections about their axes.

To edit a parameter such as *A* or *B*, double-click it with the **Arrow** tool and change the value.

1. Open the Revolution page of **Surfaces.gsp** from the **4 Other Functions** folder. The sketch shows a standard-form equation, $\frac{x^2}{A} + \frac{y^2}{B} = 1$, for a conic section in the x-y plane. The conic is centered at the origin, with its axes along the x- and y-axes. You cannot edit the equation directly, but you can change the parameters defining A and B.

The *transverse axis* is the axis that intersects the hyperbola. The other axis (the x-axis, in this case) is the *conjugate axis.*

2. Set the values of the parameters to $A = -4$ and $B = 10$. This is a hyperbola with its transverse axis along the y-axis. You can adjust the *spin, roll,* and *pitch* controls to view the hyperbola from different angles.
3. Press the *Show y Rotation* button to show the hyperbola rotating about the y-axis.

Don't use the *spin, roll,* and *pitch* controls while the surface is being painted.

4. To display the surface of revolution, press the *Paint y Surface* button. The process will take some time. You can visualize the surface more easily if you leave animation on while it is being painted. The resulting image is a trace, not a Sketchpad object, so you cannot select or manipulate it.

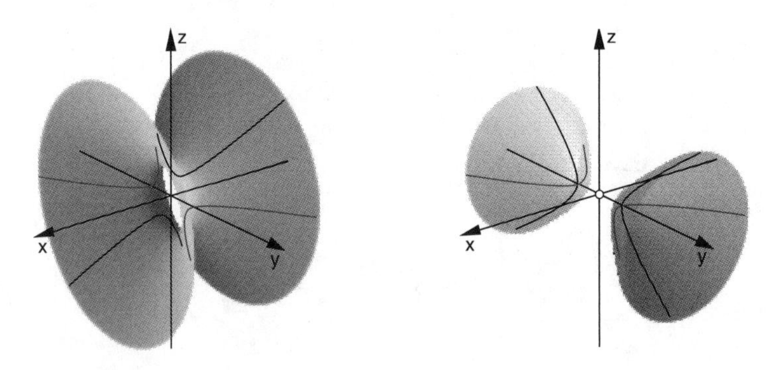

If you paint the surface a second time without erasing, the images will not appear correctly together.

5. Press the *Clear Traces* button to erase the surface. Use the *spin, roll,* and *pitch* controls to change the point of view, and paint the surface again.
6. To see what happens when the hyperbola is rotated on its conjugate axis, clear the traces and then press the *Paint x Surface* button.

Q1 Do these two surfaces intersect? Are they tangent? Describe the set of points common to both surfaces.

Q2 Experiment by changing the equation parameters. Create surfaces like the following, and record the equation and rotation you use for each.

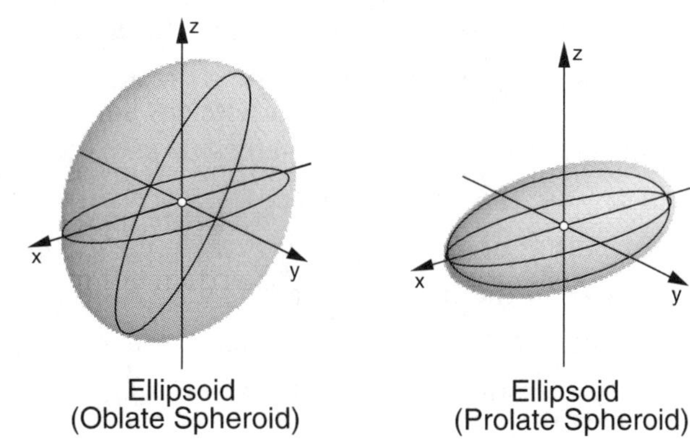

Ellipsoid (Oblate Spheroid) Ellipsoid (Prolate Spheroid)

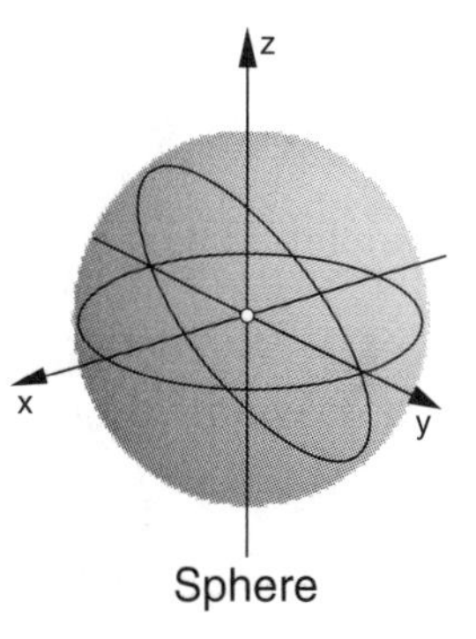

Sphere

HYPERBOLIC PARABOLOID

7. Open the Hyperbolic Paraboloid page of **Surfaces.gsp.** Near the top is an equation that defines *z* as a function of *x* and *y*. You can edit parameters *A*, *B*, and *C* to change the equation.

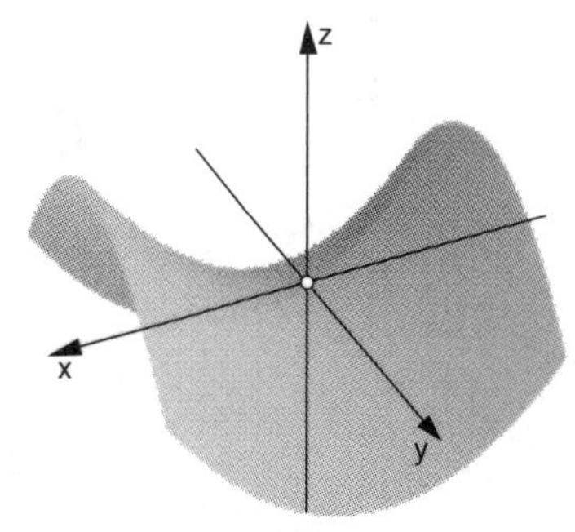

8. Press the *Paint Surface* button. Like a saddle, this surface has the interesting property of being curved both upward and downward at all points.

9. Press the *Show x-Section* button to reveal a cross-section of the surface perpendicular to the *x*-axis. Press the button again to stop the animation. You can move the cross-section manually with the slider below the equation. Similarly, you can show cross-sections perpendicular to each of the other axes.

10. Clear the traces, and use the *spin, roll,* and *pitch* controls to view the cross-sections from a different angle. Paint the surface again from this new viewpoint.

Q3 What shape is defined by each of the cross-sections perpendicular to the *x*-, *y*-, and *z*-axes?

Q4 An elliptic paraboloid has properties similar to a hyperbolic paraboloid. Its *x*- and *y*-sections are parabolas, but its *z*-sections are ellipses. Create an elliptic paraboloid by editing the parameters in this sketch. What equation did you use?

Q5 How can you create an elliptic paraboloid in which the elliptic cross-sections become circles? What should you call this surface?

EXPLORE MORE—FUNCTIONS OF TWO VARIABLES

Look again at the equation of the hyperbolic paraboloid. This equation is in the form $z = f(x, y)$. (In other words, the right side of the equation is a function of x and y.) For any real x and y, there is exactly one corresponding value for z.

You can use similar functions to describe many surfaces. Open the $f(x, y)$ page of **Surfaces.gsp.** You can use this sketch to trace a surface either in color or with a projected grid. You cannot use Sketchpad directly to create functions of two variables, but you can edit the calculation at the top of the screen to achieve the same effect. When you edit the calculation, enter x and y by clicking in the sketch on the corresponding measurements next to the calculation.

Use this sketch to create several surfaces of your own, then try to match the surface plots below.

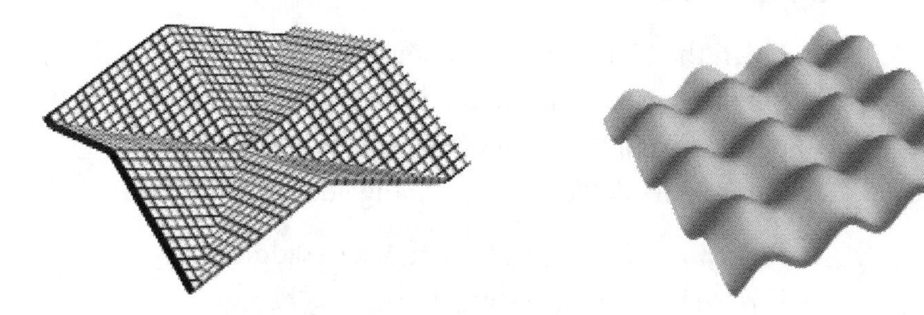

Compound Interest

It's impossible to pinpoint the moment of discovery of the constant *e*. Several branches of mathematics converged on the number from different directions. One of these was the applied mathematics of finance.

A SIMPLISTIC INVESTMENT

When an investment is compounded, the interest is paid periodically, and each time the interest itself is invested so that it can start accumulating interest as well. The more frequently the interest is paid and invested, the greater the advantage for the investor.

Depending on where you live, this interest rate may be illegal (and is certainly unlikely).

Consider a very simple investment, one dollar at 100% interest for one year. Watch what happens when it is compounded.

1. Create a new sketch and make three parameters:

$$t = 0 \text{ (the time in years)}$$

$$P = 1 \text{ (the principal)}$$

$$k = 2 \text{ (the number of compounding periods per year)}$$

2. The length of each time period will be $1/k$. Create a calculation to find the beginning of the next time period, by adding the length of one period to the original time t.

To set the precision, select the calculation and choose **Edit | Properties | Value.**

3. The interest for the first time period is P/k, so the value of the investment at the end of this period will be $P + P/k$. Express this value in factored form, and create a calculation for it. Set the precision of this calculation to hundred thousandths.

4. Plot these two points and connect them with a line segment:

(t, P) $\qquad$ $[(t + 1/k), P \cdot (1 + 1/k)]$

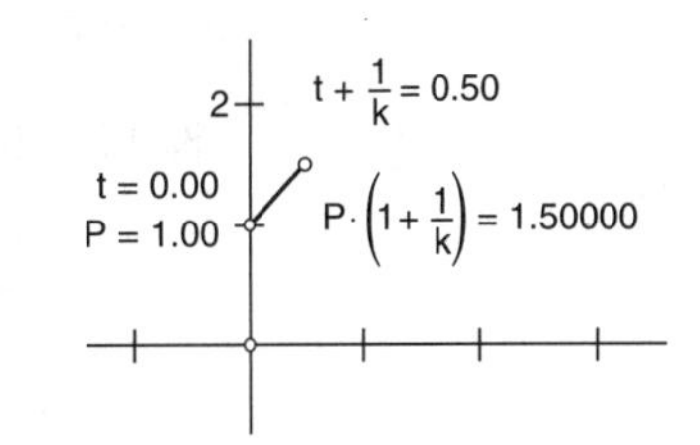

Q1 What do these points represent in terms of the investment?

5. Hide the two points, leaving only the segment to show the growth.

The calculations you just completed follow the investment for only one of the k compounding periods. You must repeat the calculations one more time to get to the end of the year.

6. Calculate $(k - 1)$. Label the calculation *depth.*

7. Select in order t, P, and *depth.* Press the Shift key and choose **Transform | Iterate To Depth.** Map the two parameters to their respective calculations as shown in the table at right.

Pre-image	First Image
t	$t + 1/k$
P	$P \cdot (1 + 1/k)$

You may need to move the table so that you can see the bottom row.

Q2 Edit parameter k to change the number of compounding periods. The bottom row of the table shows the value of the investment at the end of the year. What value must you use for k to compound the interest quarterly? What is the value of the investment after one year if it is compounded quarterly? Monthly? Weekly? Daily?

Q3 Increasing the compounding frequency always results in more money for the investor, but, as you can see, there seems to be a limit. What appears to be the limit of the value of the investment at the end of the year?

The value of this limit is known as the number e. Mathematically, you could express the limit this way:

$$\lim_{k\to\infty}\left(1+\frac{1}{k}\right)^k = e$$

A MORE REALISTIC INVESTMENT

As k grows, the investment modeled in the preceding example approaches what is called *continuous compounding.* In practice, daily compounding comes so close that the difference is negligible. Now, make some changes to your sketch and model a more realistic investment. This time, it will be \$100 at 8.5% over a term of 5 years.

8. Create these new parameters:

$r = 0.085$ (8.5% interest, displayed to the thousandth)

$term = 5$ (investment term in years)

9. Edit parameter P to make the starting principal \$100 instead of \$1. Set k to 12 for monthly compounding.

Is the graph off the screen? Use **Graph | Grid Form** to change the grid form to Rectangular and then rescale the axes.

10. The interest for the first time period is $(P \cdot r)/k$, so the value after the first time period is $P + (P \cdot r)/k$. Express this in factored form, and edit the existing calculation to match.

11. The total number of periods should now be the number of periods in a year multiplied by the number of years. Edit the *depth* calculation to achieve this. Remember to subtract one, because you've already calculated the result for one period.

Q4 What is the investment worth at the end of the term if the interest is compounded annually? Daily?

The function being modeled by this iterated calculation is the compound interest formula: $A(t) = P(1 + r/k)^{kt}$. The iteration involves repeatedly multiplying the previous result by the same factor, $(1 + r/k)$. This is an exponential function, and you write it using any base you choose. If you use e as the base, you could write it as $a \cdot e^{bt}$.

$$P(1 + r/k)^{kt} = ae^{bt}$$

Clearly, the coefficient a must equal P, but determining b is another matter.

Q5 Why must a equal P?

After setting the keyboard adjustment, you can select the parameter and press the **+** or **–** key on the keyboard to change the value of the parameter by the specified amount.

12. Create new parameter b. Use **Edit | Properties | Value** to set its precision to thousandths, and use **Edit | Properties | Parameter** to set the keyboard adjustments to 0.001.

13. Define and plot the function $A(x) = Pe^{bx}$.

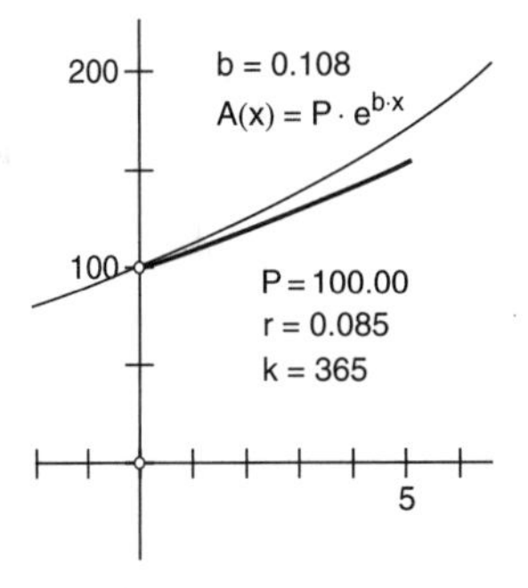

Q6 Set k to a very high number so that the iteration approximates continuous compounding. Adjust the value of parameter b so that the function graph aligns with the iterated point plot. What function models the current value of $100 compounded continuously at 8.5%?

Q7 What is the general function for the value of an investment of principal P, at interest rate r, compounded continuously for x years? Test your answer by changing the equation for $A(x)$ in the sketch. Make sure the graph of $A(x)$ always matches the iteration for different values of P and r.

Slopes of Exponential Functions

The slope of a function is its rate of change. One property that makes an exponential function so useful is the fact that it has a direct relationship with its own rate of change.

SKETCH AND INVESTIGATE

1. Open the file **Exponential Slope.gsp,** in the **4 Other Functions** folder.

To change *a* gradually, select it and press the **+** or **–** key.

To change it to a specific value, double-click it with the **Arrow** tool.

This sketch shows the function $f(x) = a^x$ and the graph $y = f(x)$. Change the base, *a*, and see how it affects the graph.

Q1 For which values of *a* is the function undefined? For which values is the slope positive, negative, or zero?

2. Press the *Show Tangent* action button to show point *A* on the graph. Line *j* is tangent to the graph at that point.

Consider the slope of the graph at point *A*. It is the same as the slope of the tangent line at that point. This is the rate at which the function is changing as *x* increases.

3. Measure the abscissa of point $A\ (x_A)$, and measure the slope of line *j*.

The slope of a function is itself a function. The function describing the slope of $f(x)$ is called the *derivative* of $f(x)$, denoted $f'(x)$. Now you will graph it.

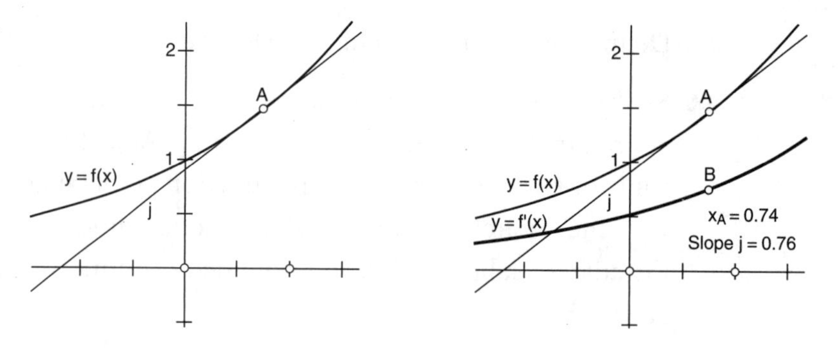

To plot point *B*, select both measurements and choose **Graph | Plot As (x,y).**

4. Plot the point $(x_A, Slope(j))$. Label the point *B*.

5. Select points *A* and *B*, and construct the locus by choosing **Construct | Locus.** Label the locus $y = f'(x)$.

Q2 Adjust the value of parameter *a* again. The graph of the derivative can fall above or below the graph of $f(x)$. For what value of *a* do the graphs coincide? What function is equal to its own derivative?

6. Compute the ratio $Slope(j)/f(x_A)$.

Drag point A along the graph. Change parameter a and drag the point again. You should see that the ratio $Slope(j)/f(x_A)$ remains constant while point A moves along the graph. Because $Slope(j) = f'(x_A)$ and the ratio is the same for any x_A, you can make this general statement:

$$\frac{f'(x)}{f(x)} = k \qquad \text{for } f(x) = a^x,\ a > 0, \text{ and some constant } k$$

If you know what k is, you should be able to solve this equation for $f'(x)$.

7. Use Sketchpad's Calculator to compute $\ln(a)$. Compare that to $Slope(j)/f(x_A)$.

Q3 Using this observation, what general statement can you make about the derivative of a^x? Solve the equation for $f'(x)$.

EXPLORE MORE

Besides e^x, there are other exponential functions that are equal to their own derivatives. Use this sketch to find them. What transformations of e^x have this property? Try horizontal and vertical translations. Try stretching it horizontally and vertically.

There is also one constant function equal to its own derivative. Try to find it.

A Sequence Approach to Logs

John Napier (1550–1617) of Scotland is credited with the invention of logarithms. Napier studied logs by making tables that placed geometric sequences and arithmetic sequences side by side. In this activity, you'll explore Napier's idea and use it to build a good approximation of a log graph.

EXAMINE A TABLE

The table at right shows a geometric sequence in the left column with common ratio r and an arithmetic sequence in the right column with common difference d. When you plot the (x, y) points, they lie on the graph of a logarithm.

x	y
r^{-3}	$-3d$
r^{-2}	$-2d$
r^{-1}	$-d$
1	0
r	d
r^2	$2d$
r^3	$3d$

Remember, $\log_c r^n = n \log_c r$.

Suppose the point (r, d) lies on the curve $y = \log_c x$ for a particular value of a constant c. This means that $d = \log_c r$.

Q1 Explain why all the other points in the table must also lie on this log curve.

CONSTRUCT A LOG CURVE

Open the sketch **Logs.gsp** in the **4 Other Functions** folder. You'll see a point F with coordinates $(1, 0)$. The sketch contains two adjustable sliders with values labeled d and r.

These were the points from the preceding table.

By plotting and connecting the points in the table, you'll obtain a good approximation of a logarithmic curve.

1. Use Sketchpad's calculator to compute the two values $x_F \cdot r$ and $y_F + d$.

This is the point (r, d).

2. Plot point G at $\left(x_F \cdot r, y_F + d\right)$. Draw a segment connecting points F and G.
3. With point F selected, choose **Transform | Iterate.** In the Iteration dialog box, designate point G as the image point by clicking on point G in the sketch.
4. Click Iterate in the dialog box to iterate the segment three times. The x-coordinates of all five points form a geometric sequence with common ratio r, and the y-coordinates form an arithmetic sequence with common difference d.
5. With the iterated image selected, press the + key on your keyboard at least 15 times to increase the number of iterations.

Now that you've created the points $(r, d), \left(r^2, 2d\right), \left(r^3, 3d\right), \left(r^4, 4d\right), \ldots$, you'll use the same technique to create the points to the left of $(1, 0)$. These points are $(1/r, -d), \left(1/r^2, -2d\right), \left(1/r^3, -3d\right), \ldots$.

6. Use Sketchpad's Calculator to compute the two values x_F/r and $y_F - d$.

7. Plot point H at $(x_F/r, y_F - d)$. Draw a segment connecting points F and H.

8. Repeat steps 3–5 with points F and H to obtain the remainder of the curve.

MANIPULATE THE CURVE

Now that your construction is complete, you'll change the values of d and r to obtain curves passing through specific points.

9. Choose **Graph | Plot Points,** and plot the point (4, 2).

You might need to increase the number of iterations if your curve shrinks.

10. Use the sliders to adjust the values of d and r so that the curve passes through (4, 2). There are many values that will work; try to find values so that the curve looks smooth.

Q2 The points on your curve (those that you created through iteration) satisfy the equation $y = \log_c x$. What is the value of c?

11. Technically, your curve is really not a curve—it's composed of many individual line segments. To see how closely your curve approximates the equation from Q2, graph the function. To do so, you'll need to remember this log identity: $\log_a b = \log(b)/\log(a)$.

To extend the function, drag the arrows on its endpoints.

12. After you've graphed the function, scroll to the right to see how well your stitched curve approximates the actual function. Notice how flat the log function is!

EXPLORE MORE

Plot the point (9, 2). Adjust the sliders so that the stitched curve passes through this point. Change the equation of the function accordingly.

Semilog Graphs

Most graphing is done on linear scales, where the distance of a point in the x or y direction is proportional to its corresponding coordinate. For some mathematical relationships, particularly ones involving both very large and very small numbers, a scale based on logarithms is more useful. This activity introduces the logarithmic scale.

THE LOGARITHMIC SCALE

On a vertical logarithmic scale, the height of a plotted point is proportional to the logarithm of its y-coordinate. Such a scale makes it possible to plot very large and very small numbers on the same scale.

30
20
10^1 10
9
8
7
6
5
4
3
2
10^0 1
0.9
0.8
0.7
0.6

1. Open the Properties page of **Semilog.gsp,** from the **4 Other Functions** folder.

On this page, only the vertical coordinate is represented. The dark horizontal line represents $y = 1$. Moving upward from there, each gridline represents one unit, up to the line $y = 10$. Above that, each gridline represents 10 units, up to the line $y = 100$. The grid density continues to be adjusted at each power of 10. Otherwise, the lines would become too closely spaced to be of any use. The same pattern continues in the downward direction.

Q1 Why is it not possible to have a negative or zero y-coordinate on a logarithmic scale?

Q2 On the left side, point A' is a dilation of point A by the indicated scale factor. What is the relationship between the y-coordinates of these two points? Drag the points and adjust the scale factor to check your answer.

Q3 Point D is formed by stacking the vertical line segments beneath points B and C. On a linear scale, y_D is the sum of y_B and y_C. What is the relationship between y_B, y_C, and y_D on a logarithmic scale? Drag the points to check your answer.

SEMILOG GRAPHS

It is often convenient to plot graphs on a *semilog* grid. This is a grid where one coordinate, usually y, is on a logarithmic scale, and the other is on a regular linear scale.

2. Open the Semilog page of the sketch.

3. Construct a free point anywhere on the grid.

4. Choose **Measure Semilog Coordinates** from the Custom Tools menu. Click the new point. Once the coordinates appear, switch back to the **Arrow** tool.

Create New Tool...
Tool Options...
Show Script View
-- This Document --
Measure Semilog Coordinates
Plot (x,y) on Semilog
Plot Function on Semilog
Function from a Line
Equation of a Line ae^(bx)
Equation of a Line a*10^bx
Equation of a Line ab^x

Q4 Drag the point slowly up and down while watching the y-coordinate. Is the y-coordinate consistent with the grid marks? Is the y-coordinate ever negative or zero?

To define the parameters, choose **Graph | New Parameter.**

5. Define parameters $x = 1$ and $y = 2$.

6. Choose **Plot (x,y) on Semilog** from the Custom Tools menu. Click on the parameters in order. A new point will be plotted.

To return the point or move it to another starting place, stop the animation and edit the parameters.

7. Select both of the parameters. Create an animation button.

Q5 The button will make both parameters increase at the same rate. Describe the motion of the point when you press the button.

EXPONENTIAL EQUATIONS

Consider an equation in the form $y = ab^x$, where $a > 0$, $b > 0$, and $b \neq 1$. You could use an equation in this form to model either exponential growth (for $b > 1$) or decay (for $0 < b < 1$). In either case, the y-intercept is a, the x-axis is an asymptote, the graph is concave up, and y is positive at any point on the graph. You will plot an exponential equation on a semilog grid.

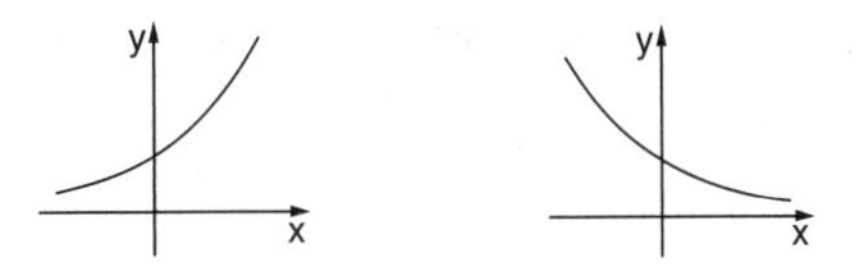

8. Open the Exponential page of the sketch **Semilog.gsp.**

9. Define two new parameters: $a = 1.50$ and $b = 3.20$.

10. Choose **Graph | New Function,** and define the function $f(x) = a \cdot b^x$.

If you plot this function normally, using the Graph menu, the plot will use the normal coordinate system, with linear axes. You must use a different method to plot on the semilog axes.

11. Choose **Plot Function on Semilog** from the Custom Tools menu. Click the function f.

12. Choose the **Arrow** tool, and experiment with changing parameters a and b.

Q6 Describe the shape of the graph. What happens to the graph when you change parameter a? What happens when you change parameter b?

Q7 In light of the observations you just made about the function plot, what is one more good reason for using a semilog grid for exponential equations?

EXPONENTIAL DECAY

Exponential equations in the form $y = ab^x$ can model quite a number of real phenomena, including population growth, compound interest, musical scales, and radioactive decay. Here is an example using radioactive decay.

Tritium $\left({}_{1}^{3}H\right)$ is a hydrogen isotope used in medical, industrial, and military applications that decays rather quickly. Suppose that 30 g of tritium is in storage. Three years later, only 25.34 g of it will remain in that form.

Now there is enough information to plot two points on the grid:

x **(time in years)**	0	3
y **(mass in grams)**	30	25.34

How long will it take for half of the tritium to be depleted?

13. Hide all of the objects that were added since opening the Exponential page of the sketch. From the Graph menu, create four new parameters having values 0, 30, 3, and 25.34.

14. Choose **Plot (x,y) on Semilog** from the Custom Tools menu. Click in order on the *x*- and *y*-coordinates to plot each of the points.

15. After plotting both points, construct a line through the two points.

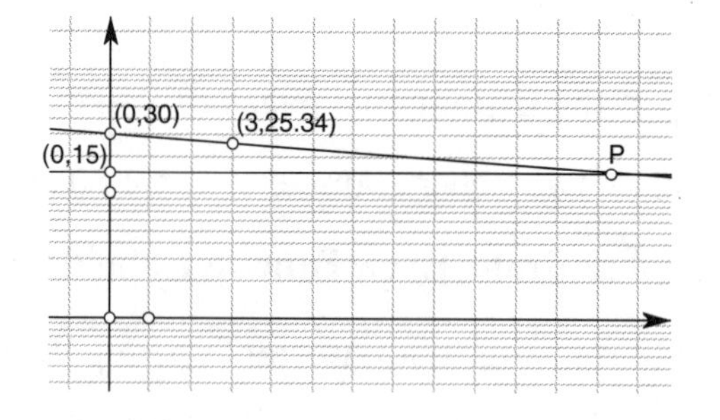

When half of the tritium is depleted, the mass of the remaining tritium will be 15 g. You'll construct a horizontal line at 15 on the vertical scale.

16. Plot the point (0, 15) by defining new parameters and using **Plot (x,y) on Semilog** again.

s that intersection point off the screen? The scale of the graph can be changed by dragging the red unit points.

17. Construct a line through this point, parallel to the x-axis. Construct the intersection point of the two lines (point P).

18. Choose **Measure Semilog Coordinates** from the Custom Tools menu and click on the intersection point P.

Q8 What are the coordinates of point P, and what do they represent? What is the half-life of tritium?

PRESENT

Although the graph is a line, its equation is not linear. The file has custom tools to represent the equation in three different forms. Use the custom tool **Equation of a Line ae^(bx)** to display the exponential decay equation. Use the custom tool **Function from a Line** to compute the function that is represented by the line.

Copy the function and the equation onto a separate page, and use them to present the same graph on a linear scale.

Log-Log Graphs

In the Semilog Graphs activity, you used a logarithmic vertical scale to simplify the plotting of certain exponential functions. In this activity, you will use a logarithmic scale on both axes. This is called a log-log graph.

SKETCH AND INVESTIGATE

1. Open page 1 of the sketch **Log-Log.gsp** in the **4 Other Functions** folder.

The sketch shows a log-log grid. The following steps refer to the dark lines as axes, but strictly speaking, they are not axes. They are the lines $x = 1$ and $y = 1$.

2. Create two new parameters a and b. For now, leave the default values $a = 1.00$ and $b = 1.00$.

3. Define a new function $f(x) = a \cdot x^b$. This type of function is called a *power function.*

Q1 With the initial settings of a and b, the definition of the function is equivalent to $f(x) = x$. How should its graph appear on the log-log grid?

4. Choose the custom tool **Plot Function on Log Grid,** and click on function f.

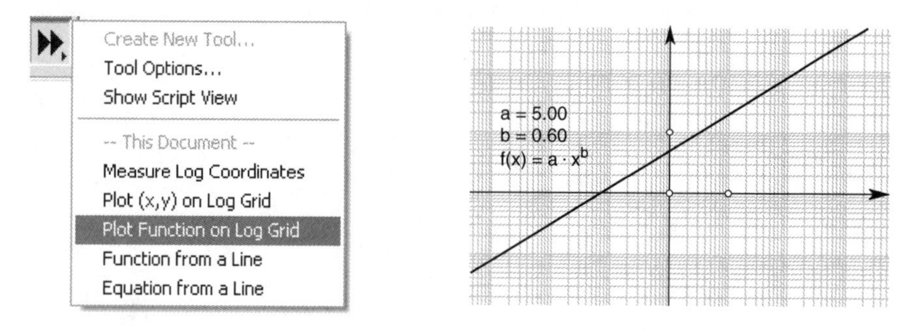

Q2 The graph you see should be a line. Experiment with editing the two parameters. Is the graph always a line? For what values of a and b is it not possible to represent the function on this grid?

The term *slope* is being used loosely here.

Q3 What are the y-intercept and slope of the line? Consider the slope the ratio of vertical change to horizontal change, as measured on a linear scale.

Q4 A point (x_0, y_0) plotted on the log-log grid is at a distance of $\log(x_0)$ linear units from the y-axis, and $\log(y_0)$ linear units from the x-axis. On the log-log grid, if a line includes point $(1, a)$ and has a "slope" of b, then by adapting the point-slope equation you get this equation:

$$\log y - \log a = b(\log x - \log 1)$$

Solve this equation for y.

APPLICATION

You have probably heard that an ant can lift ten (or more) times its own weight. Because no human can match that, it is tempting to conclude that ants have a superior physique. However, that assumes that an animal's strength is in direct proportion to its mass. Is this so?

5. Open page 2 of the sketch **Log-Log.gsp.** The grid scales have been changed and the axes hidden, but this is still the log-log grid.

These tables list the world clean-and-jerk weightlifting records recognized by the International Weightlifting Federation.

Men		Women	
Division (kg)	**Record (kg)**	**Division (kg)**	**Record (kg)**
56	168.0	48	116.5
62	182.5	53	127.5
69	197.5	58	133.0
77	210.0	63	138.0
85	218.0	69	150.0
94	232.5	75	153.0
105	242.5	75+	168.5
105+	263.0		

The parameters are from two separate data sets, so give them different colors.

6. Create a parameter corresponding to each number. Arrange them in rows and columns on the screen. Omit the bottom row of each table. The bottom row represents the super heavyweight division, which has no weight limit, hence, no independent variable.

Color the points to match the parameters.

7. Choose the Custom tool **Plot (x,y) on Log Grid.** Select the coordinate pairs in order (*division, record*).

8. For each data set, construct a line approximating the best fit with the plotted points.

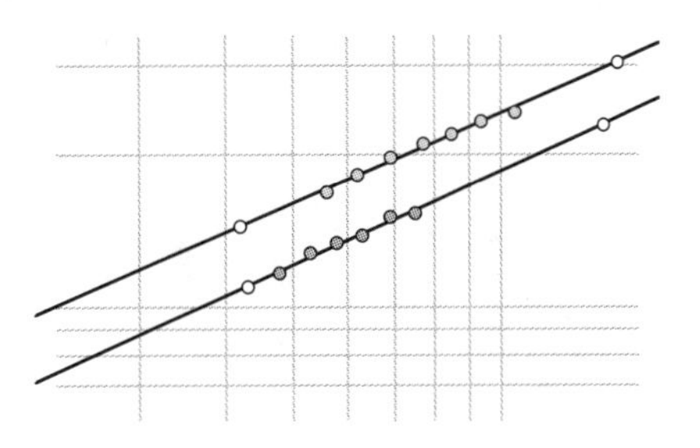

Q5 If strength were proportional to mass, what would be the slope of this line?

9. Choose the Custom tool **Equation from a Line,** and select each of the best-fit lines.

Q6 What are the equations relating the mass of a champion weightlifter to his or her record?

10. Choose the Custom tool **Function from a Line,** and again select the lines. Unlike the equations, you can use these functions in calculations.

Q7 Suppose that these athletes were scaled down to a mass of 0.1 gram, about the size of an ant. How many times his/her own weight would the athlete be able to lift? You should have two answers, one for men and one for women. Don't forget to convert the unit.

PRESENT

Select all of the data, equations, and functions, and copy them. Create a new blank page in the sketch. Paste the copied objects into the new page. Using the regular Sketchpad graphing tools, graph the data points and the functions on a linear scale.

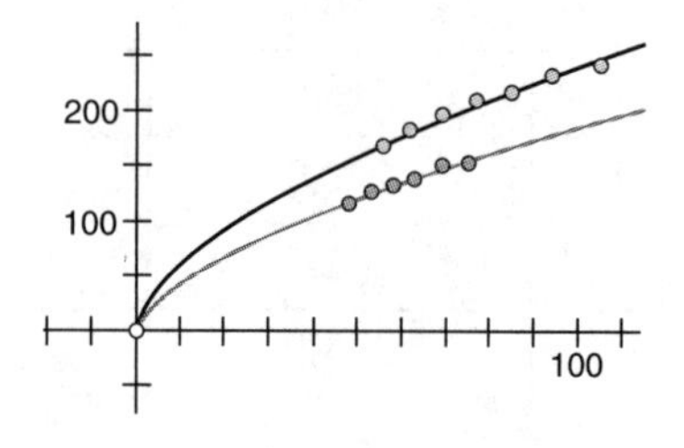

The Logistic Function

Scientists often use mathematical functions to understand natural processes. For instance, biologists and environmentalists often use mathematics to study how animal or plant populations grow larger or become smaller.

In this example, you will use Sketchpad to model a population of animals that reproduce at a certain rate, but whose numbers are limited by the available food.

DEVELOP THE EQUATION

Your mathematical model of population growth will use a function to calculate the number of animals in each generation. The input value to this function is the number of animals in the current generation, and the output value is the number in the next generation. The function includes two factors: a *growth* factor and a *limiting* factor.

For instance, if the population doubles in each generation, the value of k is 2.

Q1 For the growth factor, assume that on average each individual in one generation produces k individuals in the next generation. The value of k takes into account normal births and deaths, but not the effect of limited resources. Considering only normal growth, if the population is p in one generation, write a formula, in terms of p and k, for the population in the next generation.

If the population depends only on the growth factor, it will increase exponentially. But as the population becomes larger, food becomes scarce and population growth is limited by the lack of food. Use n to stand for the population at which the current generation doesn't have enough food to survive, and dies without producing offspring.

Taking this value n into account, a reasonable function $f(p)$ for determining the population in the next generation is $f(p) = kp(n - p)/n$. The term kp is the growth factor, and the term $(n - p)/n$ is the limiting factor.

Q2 When $p = n$, what is the value of the limiting factor? How big will the next generation be?

Q3 When p is very small compared to n, what is the approximate value of the limiting factor? How does the limiting factor affect $f(p)$?

Q4 If $n = 1{,}000{,}000$ and $k = 1.5$, find the size of the next generation for each of the following current populations: $p = 5{,}000$, $p = 50{,}000$, and $p = 500{,}000$.

You don't have to measure the population by counting individuals; you can measure it in any units you want. The mathematics will be simplest if you measure the population as a fraction of n. So, if the population is 500,000 and $n = 1{,}000{,}000$, you can define $x = p/n$ and record the population as $x = 0.5$.

Q5 If the population is measured in this way, how should you rewrite the limiting factor? Write a new formula $g(x)$ using this way of measuring population.

To construct a Sketchpad model, you will create an initial size for the population, evaluate your function, and use the result as the size of the next generation. You will apply the function repeatedly to observe the behavior of the population over time.

FIRST GENERATION

To enter the parameter *k*, choose **New Parameter** from the Calculator's Value pop-up.

1. Start with a new sketch, and choose **Graph | Plot New Function.** Plot the function $g(x) = k \cdot x \cdot (1 - x)$.
2. Choose **Graph | Plot Points,** and plot the point $(1, 1)$. Construct a diagonal segment through the origin and this plotted point. This diagonal segment represents $y = x$.

To scale the axes, drag the unit point, or drag one of the numeric labels on the axes.

3. Move the origin near the bottom left of your window, and scale the axes so that the point $(1, 0)$ is near the right edge of your sketch window.
4. Construct point *P* on the *x*-axis between 0 and 1, and measure its abscissa. This point represents the size of the initial population.
5. Calculate $g(x_P)$—the size of the next generation—and plot the point $(x_P, g(x_P))$.
6. Construct a segment connecting point *P* and the plotted point.

The calculated value of $g(x_P)$ is the output (*y*-value) of the function in the first generation, but you must use it as the input (*x*-value) for the function in the next generation. The next step provides a geometric way of turning the existing *y*-value into an *x*-value.

The *x*-value of the intersection point is the input for the next generation.

7. Construct a line through $(x_P, g(x_P))$ parallel to the *x*-axis, and construct the intersection *Q* of this parallel with the diagonal line. Hide the parallel line, and construct a segment connecting the plotted point and *Q*.

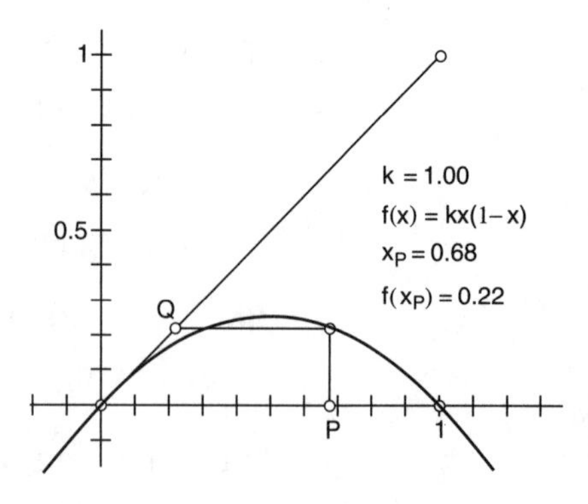

Q6 Why is x_Q equal to the value of the function at x_P? In other words, why is $x_Q = g(x_P)$ true?

This completes the construction for a single generation, because the x-value of point Q is $g(x_P)$, the population at the start of the second generation. In the next step, you will iterate this process, by using the same construction to find the population at the start of the third generation, and then the fourth generation, and so on.

NEXT GENERATIONS

8. Select point P and choose **Transform | Iterate.** In the Iterate dialog box, match point P to point Q by clicking on point Q in the sketch. In the Structure pop-up, make sure that Tabulate Iterated Values is checked. Click the iterate button in the dialog box to show the first three steps of the iteration.

You can also use **Edit | Properties | Iteration** to change the number of iterations.

9. Increase the number of iterations to 20 by pressing the + key on the keyboard.

In the questions that follow, you are asked to observe the long-term behavior of the population under various circumstances. You can observe the long-term behavior of the population by looking at the iterated segments or by observing the values near the bottom of the table. Be sure to look at the long-term behavior, ignoring the first few generations.

You may need to increase the number of iterations to answer this question.

Q7 Move point P to about 0.5. Does the population ever stabilize? If so, at what value does it stabilize?

Q8 Drag point P back and forth between 0 and 1, and observe the behavior. Does the long-term behavior of the population depend on the initial position of P? If so, in what way does it depend on P? If not, explain why the long-term behavior is the same no matter where P is.

Q9 Change the value of k to 2.5. What's the long-term behavior of the population now? Does it depend on the initial position of point P?

Q10 Change the value of k to 3.1. What's the long-term behavior of the population now? Explain this behavior. Does it depend on the initial position of point P?

10. Change the properties of parameter k so that the keyboard adjustment is 0.01.

Q11 Select parameter k and use the + key on your keyboard to gradually increase the value of k from 2.5 to 4.0. Record your observations. For particular values of k, drag P to look at the sensitivity of the long-term population to the initial population. Also observe the stability of the long-term population as the value of k is changed slightly.

EXPLORE MORE

To actually derive one of these forms of the logistic equation from the other requires calculus.

The function used in this activity produces results at discrete intervals of time, generation by generation. This discrete form is sometimes called a *logistic map*. You can also express the logistic function continuously, in a form that gives the size of the population as a function of time. Here's a form that uses k in a similar way:

$$p(t) = \frac{k - 1}{k + c \cdot e^{-(k-1)t}}$$

Plot this function, using parameters for c and k. Experiment with different values of the parameters, and observe how the behavior of this form of the logistic function relates to the behavior of the Sketchpad model you have built. What similarities do you observe? How can you account for the differences?

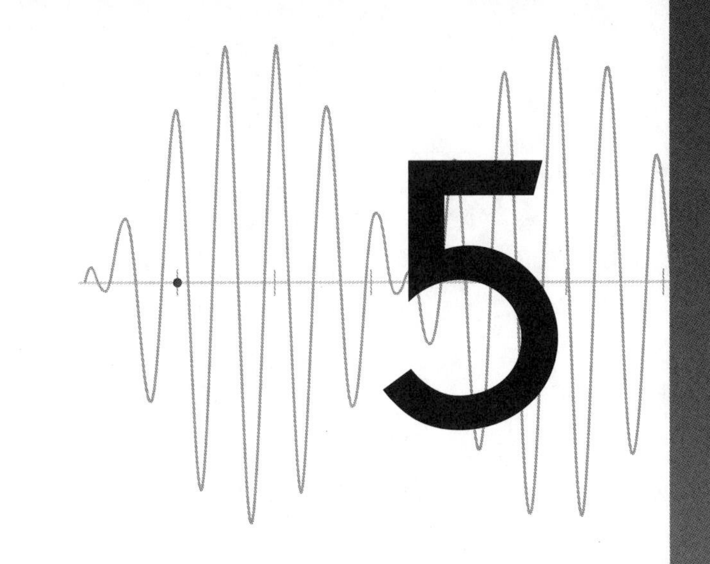

Data and Probability

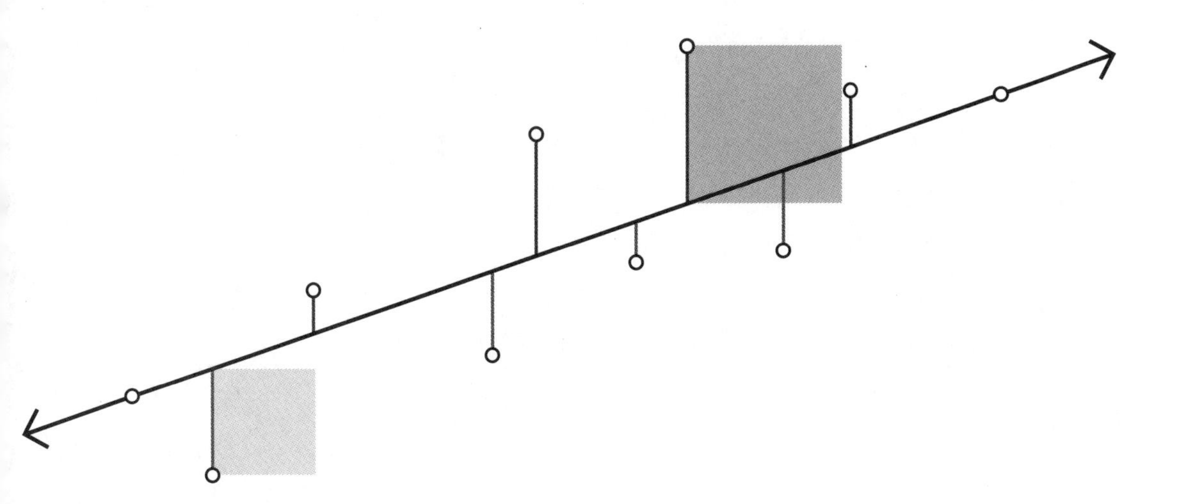

Linear Regression

When analyzing data involving two variables, you often want to find a straight line that fits the data well. You may have performed this task graphically by looking at data points and drawing a line that comes close to all the points. In this activity you will do this operation mathematically, and you will analyze the results.

SKETCH AND INVESTIGATE

1. Open the document **5 Data and Probability | Linear Regression.gsp.**

The sketch contains eight numbered data points that you can drag wherever you want and a regression line *QR*. The *residual* of a data point is its vertical distance from the line. The residuals for all of the points are displayed in the left side of the sketch.

2. Drag points *Q* and *R* to make the line come close to all the data points. Press the *Show Residual Segments* button.

Q1 What do the lengths of these segments represent?

Q2 Some segments are red and some are blue. What do the colors indicate?

You may think that the best fit will result from minimizing the total of all the residuals. You will test this idea in the next few steps.

3. Calculate the sum of the eight residuals. Adjust the line to minimize this calculation.

Q3 What was the smallest value you were able to get for the sum?

Q4 Can you adjust the line to fit the data badly, but still minimize the residual sum?

A bad fit may have a small sum because the residuals may be positive or negative.

Q5 Would it help to square the residuals before adding them? Why or why not?

This method is called a least squares analysis.

4. Calculate the sum of the squares of the residuals $\left(r_1^2 + r_2^2 + \ldots\right)$. Click the *Show Squares* button.

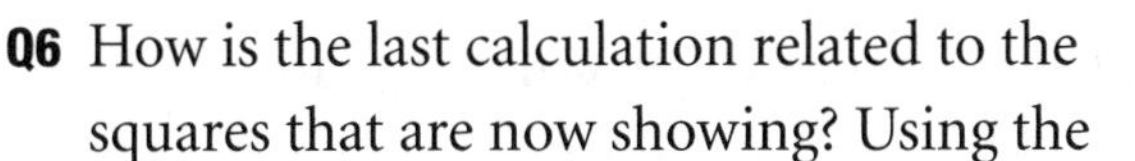

The *Seek Least Squares* button simulates a trial-and-error process for minimizing the squares, but try doing it by yourself a few times.

Q6 How is the last calculation related to the squares that are now showing? Using the geometric representation as a guide, adjust the regression line to minimize the sum of the squares. Drag the points, and try several arrangements.

Q7 When the data are arranged as in the plot at right, how would you expect the best-fit line to be oriented?

5. Arrange the data points to roughly approximate the picture. Then drag Q and R to minimize the sum of the squares.

Q8 Describe the line that actually fits the data best. Does the result surprise you?

To minimize the residuals, you could try adding absolute values rather than squares.

6. Compute the sum of the absolute values of the residuals ($|r_1| + |r_2| + \ldots$). Drag Q and R to minimize this sum.

The *Seek Least* $|r|$ button will seek to minimize the sum of the absolute values.

Q9 How different is this line from the one that minimizes the sum of the squares? Repeat for several arrangements of the data points, and record your conclusions. Why do you suppose statisticians normally minimize the squares rather than the absolute value?

ENTER DATA

7. Press the *Show Sleep Data* button. The data set that appears represents average sleep per night (x, in hours) and final exam test scores (y) of a small group of students.

8. Press the *Set all 8 Points* button to send the data points to these positions. (You could drag the points, but pressing the button is faster.)

To find the equation of a line, select it, and choose **Equation** from the Measure menu.

Q10 Based on a least squares analysis, what is the equation of the regression line?

EXPLORE MORE

Many calculators can compute the equation of the least squares regression line for a set of data. Enter the sleep data into a calculator, and find the equation in the form $y = ax + b$. In the sketch, enter the appropriate values for parameters a and b, and click the button below them. Can you improve on this fit?

You can enter data from other sources, but only if there are no more than eight data pairs. To enter fewer than eight, you must eliminate the residuals of the unused points. To do so, select an unused point and line PQ, and choose **Edit | Merge Point To Line.** You can then hide the point, but do not delete it.

Wait for a Date

You and a friend arrange for a lunch date next week between 12:00 and 1:00 in the afternoon. A week later, however, neither of you remembers the exact meeting time. As a result, each of you arrives at a random time between 12:00 and 1:00 and waits exactly 10 minutes for the other person. When the 10 minutes have passed, each of you leaves if the other person has not come.

What is the probability that the two of you will meet?

MODEL IN ONE DIMENSION

1. Open the sketch **Wait for a Date.gsp** in the **5 Data and Probability** folder. The times when you and your friend arrive are represented as two points, *A* and *B*, at random locations along a timeline. In their initial locations, *A* and *B* arrive 6 minutes apart, well within your 10-minute limit.

2. Press the *Do A and B meet?* button several times. Doing so moves points *A* and *B* to new, random locations along the segment. Notice how the appearance of the stick figures changes depending on the time interval separating them.

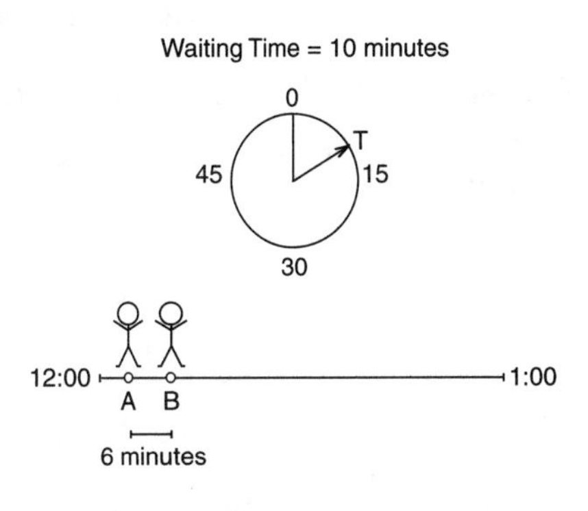

3. Drag point *T* to change the amount of time you and your friend will wait. Now run the simulation again several times.

Q1 How can you change the waiting time to make it more likely that you and your friend will meet?

MODEL IN TWO DIMENSIONS

Viewing a one-dimensional Sketchpad simulation helps you to understand the problem, but it does not allow you to compute the exact probability that you and your friend will meet. A shift in perspective from a one-dimensional to a two-dimensional model makes a big difference, as you will soon discover.

4. Open the second page of the sketch. As before, points *A* and *B* are on a horizontal timeline, but now there is a vertical timeline that shows when *B* arrives.

Q2 Point P is at the intersection of lines through A and B that are perpendicular to the two axes. The following picture shows four possible locations of P. For which two locations does it appear that you and your friend meet?

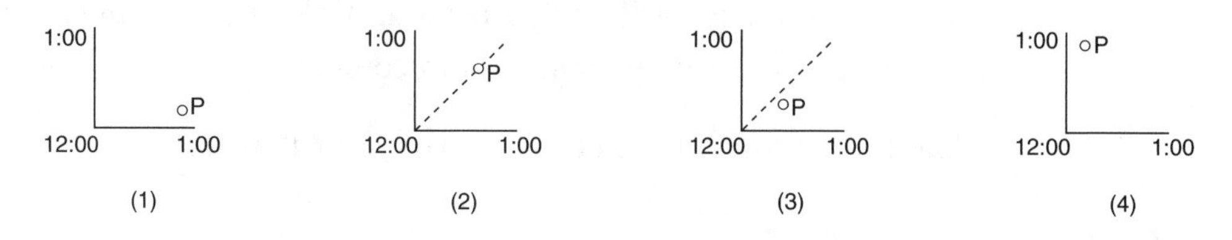

5. Press the *Do A and B meet?* button. Doing so moves points A and B to new, random locations along the segment. When A and B meet, point P leaves a green trace on-screen. When the two do not meet, the trace is red.

6. To speed up the process, choose **Display | Show Motion Controller,** and click on the upward-pointing arrow to increase the speed.

To clear the screen, choose **Erase Traces** from the Display menu.

Q3 Run the simulation for a while. Describe the emerging pattern of red and green points.

Q4 Draw the set of all points P for which you and your friend arrive at the exact same time. Describe this set of points.

Q5 Draw the set of all points P for which your friend (point B) arrives exactly 10 minutes after you (point A). Describe this set of points.

Q6 Draw the set of all points P for which you arrive exactly 10 minutes after your friend. Describe this set of points.

Use your answers to Q4–Q6 to help with the next two questions.

Q7 What portion of the square formed by the two axes is filled with red points? What portion is filled with green points?

Q8 What is the probability that you and your friend will meet?

EXPLORE MORE

Q9 Drag point T to change the waiting time. Now run the simulation again. What is the probability that you and your friend will meet?

Q10 If you and your friend are willing to wait t minutes for each other, what is the probability you'll meet?

Q11 How do things change if you are willing to wait for 15 minutes, but your friend is willing to wait for only 5 minutes?

Fitting Polynomial Functions

Consider two arbitrary points on the x-y coordinate plane with different x-coordinates. You can define a linear function of x whose graph passes through both points. This is a simple case of a more general concept. Given n points on the plane, no two of them with the same x-coordinate, there is exactly one polynomial function of x, of degree $n - 1$ or lower, whose graph includes all of the given points.

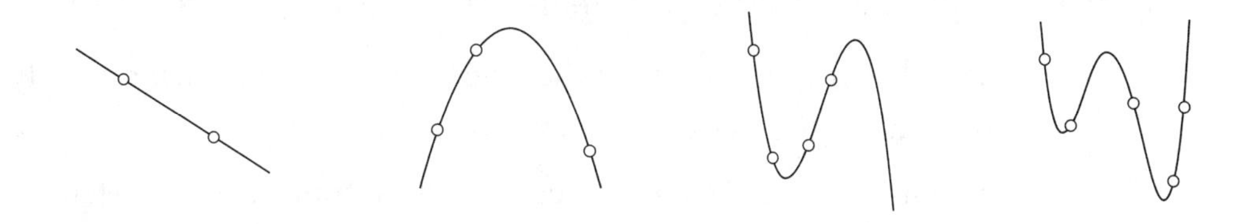

In this activity you will derive a function to fit three points. Since there are three points, your function will be of degree 2 or less.

Q1 Why does the problem require that no two of the points have the same x-value?

SKETCH AND INVESTIGATE

Use numbers as the point labels because the labels will become subscripts later.

1. Open the sketch **Polynomial Fit.gsp** from the **5 Data and Probability** folder. Construct three free points labeled *1, 2,* and *3*. Move the points near the right side of the screen; you will need some room for the calculations.

You are using three points in this case, so the objective is to find the coefficients for the 2nd-degree polynomial function $f(x) = a_1 + a_2x + a_3x^2$.

2. For each of the three points, measure its x- and y- coordinates, and compute x^2. You now have enough information to form a system of three equations.

$$a_1 + a_2x_1 + a_3x_1^{\,2} = y_1$$
$$a_1 + a_2x_2 + a_3x_2^{\,2} = y_2$$
$$a_1 + a_2x_3 + a_3x_3^{\,2} = y_3$$

The variables a_1, a_2, and a_3 are written as coefficients here, but they are in fact the only unknowns in the system. You can express this system in matrix form:

To see that these are equivalent, think of the two equations as $M \cdot a = y$ and $M^{-1} \cdot y = a$.

$$\begin{bmatrix} 1 & x_1 & x_1^2 \\ 1 & x_2 & x_2^2 \\ 1 & x_3 & x_3^2 \end{bmatrix} \begin{bmatrix} a_1 \\ a_2 \\ a_3 \end{bmatrix} = \begin{bmatrix} y_1 \\ y_2 \\ y_3 \end{bmatrix} \text{ or, equivalently, } \begin{bmatrix} 1 & x_1 & x_1^2 \\ 1 & x_2 & x_2^2 \\ 1 & x_3 & x_3^2 \end{bmatrix}^{-1} \begin{bmatrix} y_1 \\ y_2 \\ y_3 \end{bmatrix} = \begin{bmatrix} a_1 \\ a_2 \\ a_3 \end{bmatrix}$$

You can find the unknowns with the help of an inverse matrix.

Another way to enter the left column would be to compute 3 values for x^0.

3. Arrange the measurements to form the 3×3 matrix on screen. You will need to create parameters for the 1's in the left column.

4. Choose the custom tool **3D Inverse.** Click the tool on the matrix elements in the order shown at right. Start with the left column, from top to bottom, then the middle column, then the right. The elements of the inverse matrix will appear in the same order, but as a single column. Rearrange them to form the inverse matrix.

1.00	$x_1 = -2.38$	$x_1^2 = 5.67$
1.00	$x_2 = 1.03$	$x_2^2 = 1.06$
1.00	$x_3 = 1.67$	$x_3^2 = 2.78$

Now that you've computed the inverse matrix, you must multiply this inverse matrix by vector **y** in order to obtain vector **a.**

5. Choose the **3D Matrix*Vector** custom tool. Select the elements of the inverse matrix, column by column, again. Then select the elements of vector **y** in order: y_1, y_2, y_3. The elements of vector **a** will appear in order: a_1, a_2, a_3. Change their labels appropriately.

6. Define and plot the function $f(x) = a_1 + a_2x + a_3x^2$.

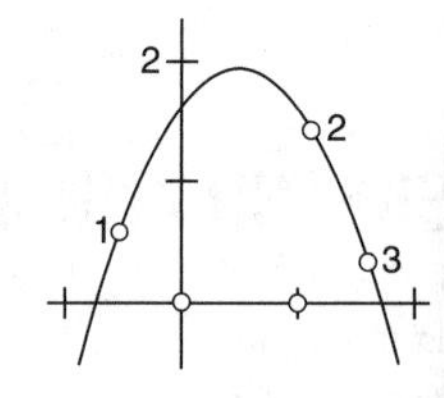

Q2 The introduction to this activity states that there is always a solution of degree $n - 1$ or less. Under what conditions will this method result in a polynomial function of degree less than $n - 1$?

7. The polynomial that you just defined and graphed will lose a degree when a_3 is zero. Drag points 1, 2, and 3 to see when that happens.

Q3 Are there any polynomial functions of degree n or greater that will fit all of the given points? How many?

Q4 What happens to the inverse matrix when two of the points have the same x-coordinate?

PRESENT

See **Help | Advanced Topics | Advanced Text Topics | Merging Text with a Custom Template** to learn about text templates.

Although your curve fits all of the points, the function itself is hard to read. You can change it so that you can clearly see the coefficients as the points are dragged. Using the **Text** tool, create this template:

$$=f(x) = \{1\} + (\{2\})x + (\{3\})x^2$$

Select, in order, the above template and a_1, a_2, and a_3. From the Edit menu, choose **Merge Text.**

EXPLORE MORE

This sketch includes custom tools that will accommodate matrices up to dimension 5. Add new pages to the sketch and use the **Custom** tools to fit a function to four points and then to five points.

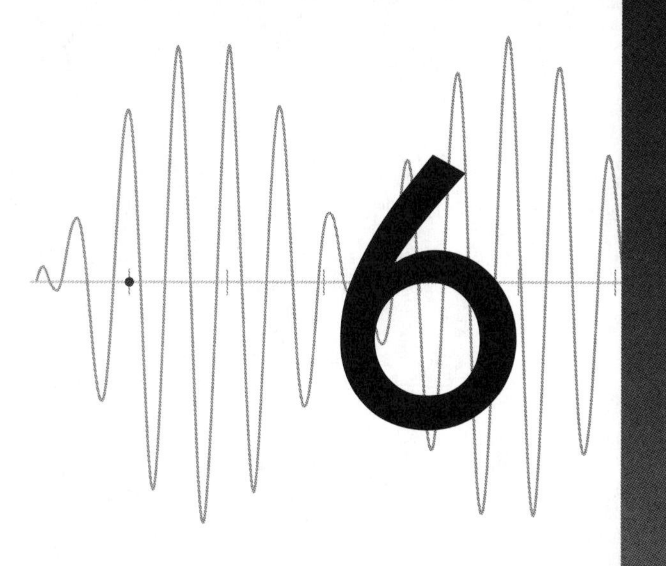

6

Vectors and Matrices

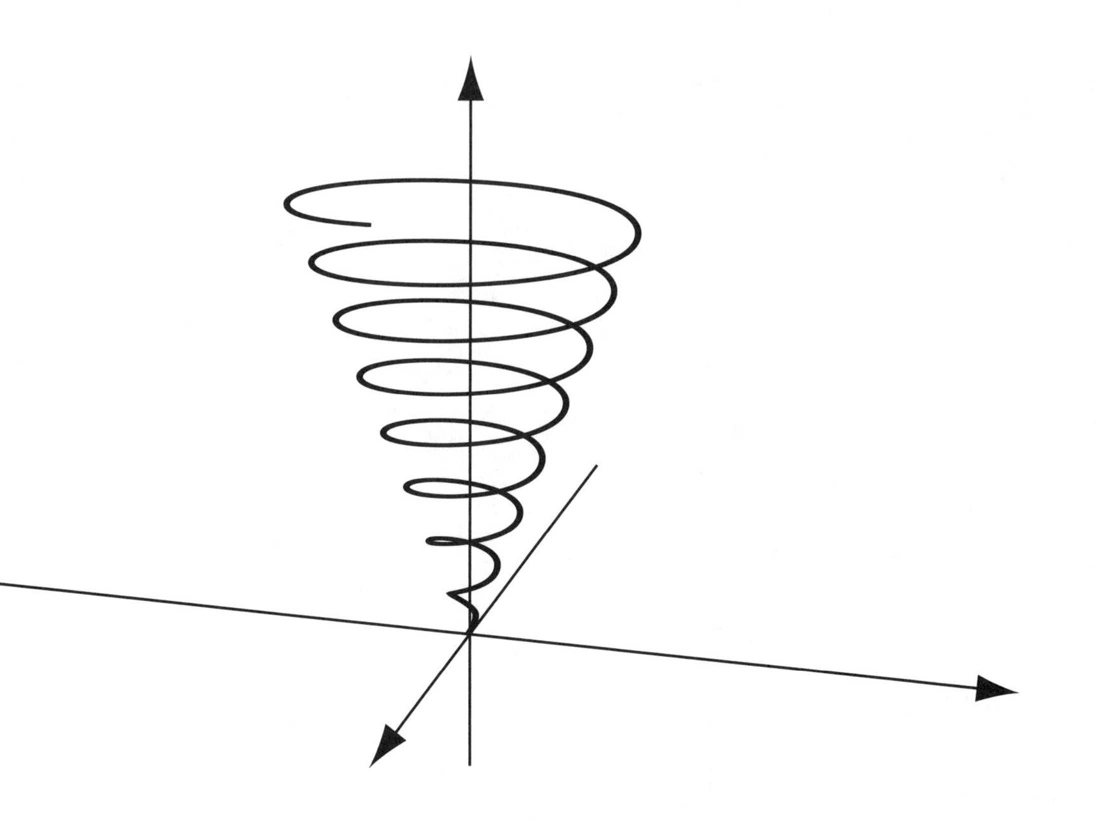

Matrix Transformations

You can use matrix multiplication to transform coordinates very efficiently. This procedure is useful in many fields, including computer programs such as The Geometer's Sketchpad itself. In this activity you will use matrices to duplicate some of Sketchpad's transformations.

TRANSFORM A POLYGON

1. Open the Simple page of **Matrix Transformations.gsp** in the **6 Vectors and Matrices** folder.

A vector can be represented as a 2×1 matrix whose elements are the horizontal and vertical components of the vector.

The sketch has a 2×2 matrix. Next to the matrix is a vector (in the form of a 2×1 matrix) representing the coordinates of point P, an independent point.

$$\begin{bmatrix} -1 & 0 \\ 0 & -1 \end{bmatrix} \begin{bmatrix} x_p \\ y_p \end{bmatrix}$$

Q1 By hand, multiply the matrix by the vector (x_P, y_P). Describe the resulting vector geometrically in terms of point P and the x-y coordinate system.

2. Choose the **2D Matrix*Vector** custom tool. Click on the elements of the matrix, column by column, from top to bottom, and then the elements of the vector.

Create New Tool...
Tool Options...
Show Script View
-- This Document --
2D Matrix * Vector

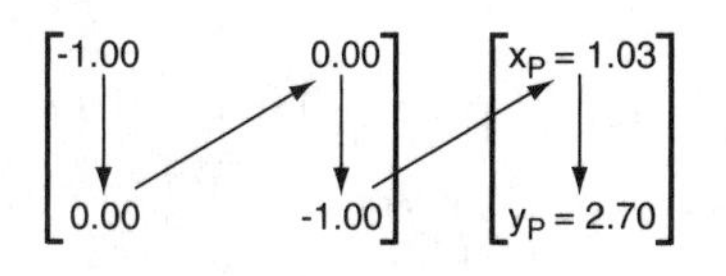

3. The result is a pair of coordinates. Choose the **Arrow** tool, select the coordinates, and plot them on the x-y grid. Label the new point P'.

You might want to apply this transformation to some object other than a point. A 2×1 matrix can be used to represent a vector or a point but cannot represent any other object. To work around this problem, put the point on a path.

To construct a polygon interior, construct the vertices, select them, and choose **Construct | Polygon Interior.**

4. Construct a polygon interior, using at least five points.

5. Select P and the interior. Choose **Edit | Merge Point To Polygon Interior.**

6. Drag point P to see that its path is the boundary of the polygon. Select points P and P', and construct the locus of P' as P moves along the polygon boundary.

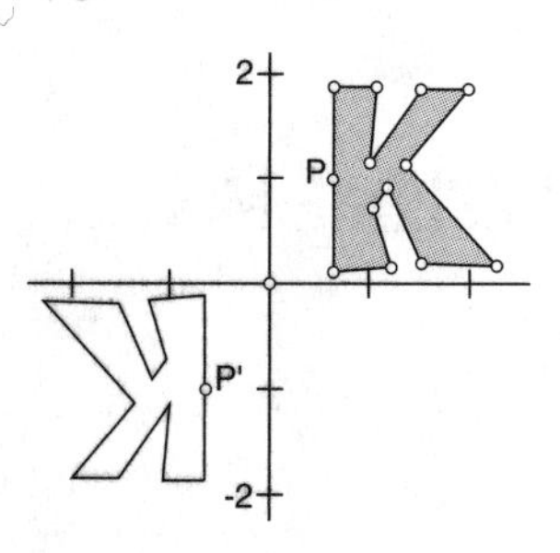

Q2 Apply each transformation matrix below to the coordinates of point P. For each matrix, predict the image of the transformation, try it out in the sketch, and describe it geometrically.

Ratio r, angle θ, and their controls are on the left side of the sketch.

You can create a new matrix for each transformation, but it will be easier to edit the existing matrix. To insert a calculation such as $\cos\theta$ in place of a matrix element, select the element and choose **Edit | Edit Parameter.**

$\mathbf{A} = \begin{bmatrix} -1 & 0 \\ 0 & 1 \end{bmatrix}$	$\mathbf{B} = \begin{bmatrix} 1 & 0 \\ 0 & -1 \end{bmatrix}$	$\mathbf{C} = \begin{bmatrix} 0 & 1 \\ 1 & 0 \end{bmatrix}$	$\mathbf{D} = \begin{bmatrix} r & 0 \\ 0 & r \end{bmatrix}$
$\mathbf{E} = \begin{bmatrix} r & 0 \\ 0 & r \end{bmatrix}$	$\mathbf{F} = \begin{bmatrix} 1 & 0 \\ 0 & r \end{bmatrix}$	$\mathbf{G} = \begin{bmatrix} \cos\theta & -\sin\theta \\ \sin\theta & \cos\theta \end{bmatrix}$	$\mathbf{H} = \begin{bmatrix} \cos\theta & -\sin\theta \\ \sin\theta & \cos\theta \end{bmatrix}$

COMBINE TRANSFORMATIONS

You can also do translation using matrices, but that technique must wait for another activity.

These transformation matrices may not be as flexible as you would like. For example, you can use such a matrix to rotate about the origin, but Sketchpad's built-in tools can rotate an object about any point, not just the origin. You can combine rotation with translation to do the same thing. To translate, you will add and subtract coordinates.

7. Open the Combo page of **Matrix Transformations.gsp.**

This is the same as the first page, except for the addition of another point, Q. The matrix is an identity matrix waiting to be edited. Your objective is to rotate point P about point Q.

There is no need to actually plot point A, but doing so may help you understand the process.

8. Subtract the coordinates of point Q from P. Label the new coordinates x_A and y_A. This effectively translates P by vector $\overrightarrow{QO}$ to point A.

9. Edit the matrix to create a clockwise rotation by angle θ. Use the **2D Matrix*Vector** custom tool to multiply this matrix by the coordinates of point A. Label the results x_B and y_B.

10. Add the coordinates of Q to those of B. Label the sums x_P' and y_P'. Plot these coordinates as point P'. As you did before, merge point P to a path, and construct a locus.

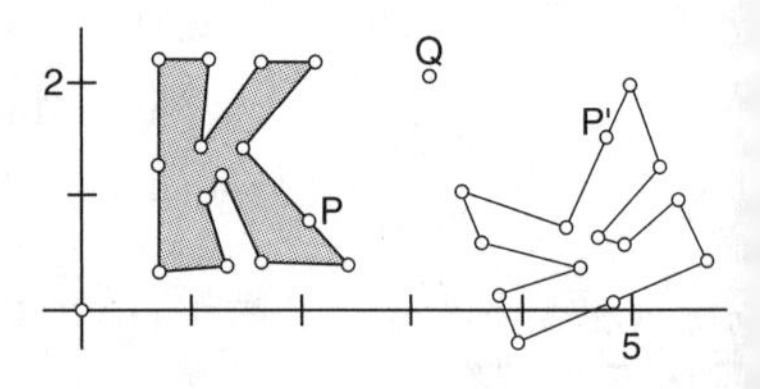

Q3 How can you edit this matrix to dilate P, using r as the scale factor and Q as the center of dilation? Check your answer by doing the dilation in the sketch.

EXPLORE MORE

Multiply two or more of the matrices, and predict the transformation that will result from the matrix product. For example, referring to the list of matrices presented earlier, what is the transformation resulting from matrix product **GE**? Is it the same as **EG**?

Vector Operations

You can use vectors to model concepts that are not at all geometric in nature, but it is often helpful to keep the geometric connection in mind. In this activity you will explore vector projections and dot products, first geometrically, then with an emphasis on calculation.

GEOMETRY

1. Open the Geometry page of **Vector Operations.gsp** in the **6 Vectors and Matrices** folder. The sketch contains two vectors in the *x*-*y* plane.

The vector components were calculated from the point coordinates.

You can translate a vector by dragging its initial point. To change the magnitude or direction of a vector, click the appropriate action button and drag the terminal point. The point coordinates and vector components change as you modify the vector.

Q1 Just to be sure, compute the components of vector **a** yourself from the point coordinates. What formulas did you use?

Q2 What effect does translating a vector have on the vector components? Verify your answer by dragging the initial point of a vector.

To choose a custom tool, press and hold the **Custom** tool icon at the bottom of the Toolbox.

2. You can compute the dot product of two vectors from the vector components. If you know a formula for this, use it. Otherwise, choose the **Dot Product** custom tool, and click on a_1, a_2, b_1, and b_2.

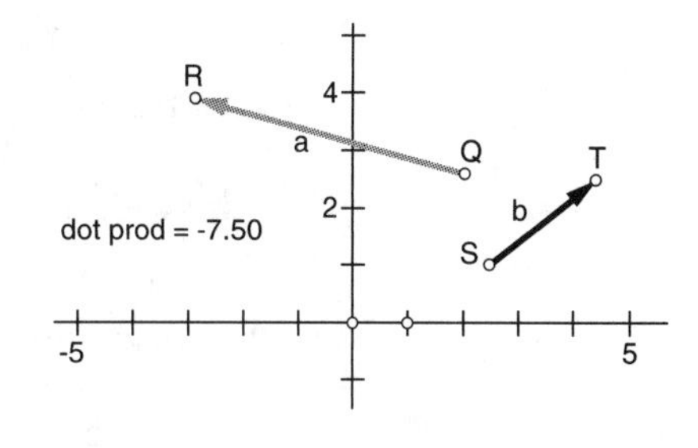

Q3 What happens to the dot product when you translate a vector?

To change a vector's magnitude or direction, press either the *Change Magnitude* or *Change Direction* button and then drag the terminal point.

Q4 What happens to the dot product when you change the magnitude of a vector?

Q5 What is the relationship of the vector directions when the dot product has its maximum value? When the dot product is zero? When the dot product has its minimum (negative) value?

3. Click the *Show Projection* button to show the projection of vector **a** onto **b.**

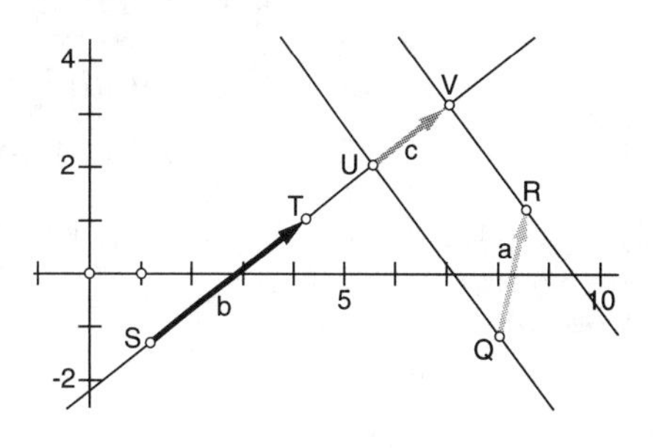

Q6 You can translate, change the direction of, or change the magnitude of either vector. Predict which of these actions will change vector **c.** Make a copy of the following table, and mark each cell with a *Y* if you predict that it will change **c** and an *N* if it will not. Check your predictions.

	Translate	Change Direction	Change Magnitude
a			
b			

CALCULATIONS

In this section you will calculate first the projection of vector **b** on vector **a,** and then the dot product **a · b.**

In the previous section you observed that translating vectors has no effect on a dot product or a vector projection. You can simplify the visualization and the calculations by translating the initial points of both vectors to the origin. After you translate the initial point to the origin, the components of a vector are simply the coordinates of the terminal point.

On the Calculations page of **Vector Operations.gsp,** two vectors (**a** and **b**) have their initial points at the origin and have terminal points *A* and *B*. You will begin by using vector components to compute the projection of **b** on **a.**

4. Measure the magnitude of vector **a** by choosing the custom tool **Vector Magnitude** and clicking the components of **a.** Label the calculation $|\mathbf{a}|$.

A vector is a quantity that has both magnitude and direction; a scalar is a quantity that has magnitude only.

The *unit vector* of **a** (denoted $\hat{\mathbf{a}}$) has the same direction as **a,** but magnitude 1. To compute it, divide the components of **a** by the vector magnitude: $\hat{\mathbf{a}} = \mathbf{a}/|\mathbf{a}|$. You will do the actual calculation by multiplying the scalar quantity $(1/|\mathbf{a}|)$ by the vector **a.**

5. Use the Calculator to compute the scalar $1/|\mathbf{a}|$.
6. To multiply this scalar by vector **a,** choose the custom tool **Scalar*Vector** and click on (in order) $1/|\mathbf{a}|$, the *x*-component of **a,** and the *y*-component of **a.**

Select the components and choose **Graph | Plot As (x, y).**

7. Plot the resulting components as a vector, and label the terminal point *A^*.

Q7 Where does point $A^\wedge$ end up? How far is it from the origin?

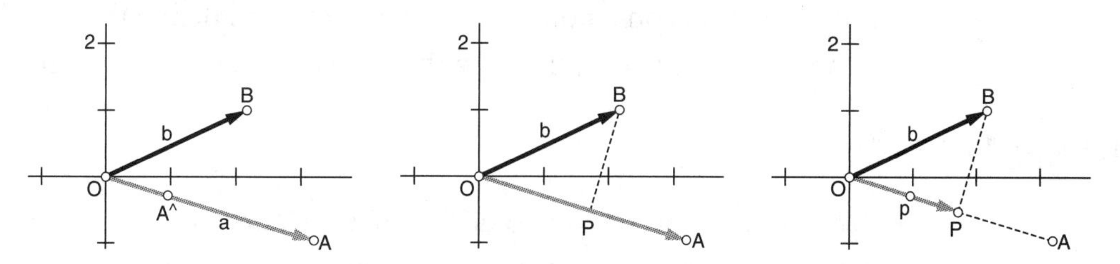

This unit vector $A^\wedge$ establishes the direction of the projection **p.** You must also determine the magnitude of **p.**

In triangle *OBP* in the figure at far right, $\cos\angle AOB = \frac{|\mathbf{p}|}{|\mathbf{b}|}$, so $|\mathbf{p}| = |\mathbf{b}|\cos(\angle AOB)$.

8. Use the **Vector Magnitude** tool to calculate the magnitude $|\mathbf{b}|$.

9. Measure $\angle AOB$, and then calculate $|\mathbf{b}|\cos(\angle AOB)$. This calculation is the magnitude of **p,** so label it $|\mathbf{p}|$.

10. You now have both the magnitude and direction of **p.** To plot **p,** multiply the magnitude $|\mathbf{p}|$ (which is a scalar) by the unit vector **â.** Use the **Scalar*Vector** tool to do the multiplication.

11. Use the components of **p** to plot the terminal point *P* of the vector. Use the **Arrowhead** tool to draw the vector itself, and label it **p.** Make **a** dashed so that you can see **p** more easily.

Q8 Drag *A* and *B* around, and observe the behavior of **p.** What happens to **p** when $\angle AOB$ is obtuse?

Q9 The vectors **â** and **p** actually may have opposite directions. If that is the case, why do these calculations still work? Shouldn't the formula $|\mathbf{p}|\hat{\mathbf{a}}$ result in a vector that is turned in the wrong direction?

12. The dot product of two vectors **a** and **b** is defined as $|\mathbf{a}||\mathbf{b}|\cos(\angle AOB)$. You have already calculated $|\mathbf{b}|\cos(\angle AOB)$, because that is $|\mathbf{p}|$. Use the Calculator to compute the dot product from your existing calculations for $|\mathbf{a}|$ and $|\mathbf{b}|\cos(\angle AOB)$.

13. Use the **Dot Product** tool on the components of **a** and **b,** and compare the result with the value you calculated in step 12.

14. Use the Calculator to compute $a_1b_1 + a_2b_2$. Compare this result with the dot products you calculated in steps 12 and 13.

Q10 Drag *A* and *B* to change the two vectors. Do the various calculations of the dot product agree as you change the vectors?

Matrix Products

In this activity you will build a single matrix to perform more complicated transformations on a point in the plane. You will define transformations, not just with respect to the origin, but with respect to other arbitrary points.

TRANSLATION

Matrix operations handle rotation, dilation, and even reflection very well, but translation presents a problem: you cannot create a translation with a 2×2 matrix. You can solve this problem by augmenting the matrix and the vector, as shown here:

$$\begin{bmatrix} 1 & 0 & a \\ 0 & 1 & b \\ 0 & 0 & 1 \end{bmatrix} \cdot \begin{bmatrix} x \\ y \\ 1 \end{bmatrix} = \begin{bmatrix} x + a \\ y + b \\ 1 \end{bmatrix}$$

The third element of the vector is always a 1. A matrix in this form translates the vector by a horizontally and by b vertically. The third element of the resulting vector is always 1, which you can disregard.

1. Open the Translation page of **Matrix Product.gsp,** in the **6 Vectors and Matrices** folder. The sketch contains a translation matrix and the augmented vector representing point P in the x-y plane.
2. To translate point P, choose the **3D Matrix*Vector** custom tool, and click first on the matrix elements, column by column, from the top down, and then on the vector elements.
3. The result should be three numbers representing the translated image of P. Plot the first two as (x, y), and label the point P'.

You can change the translation by editing the top two elements in the third column of the matrix.

Now use two points to define a translation, just as Sketchpad does.

4. Press the *Show Q & R* button. This reveals two more points and their coordinates.

To change a parameter into a calculation, select the parameter and choose **Edit | Edit Parameter.**

Q1 Using the coordinates of points Q and R, edit the top two elements of the third column to create a translation by vector QR. Record the new matrix definition on paper. Drag points P, Q, and R to test the results.

ROTATION / DILATION

You can define 2 × 2 matrices that rotate, dilate, or reflect. You can perform any of these transformations with a 3 × 3 matrix by embedding the 2 × 2 matrix in the upper left part of a 3 × 3 matrix. Here are some examples:

Notice that all of these examples have a familiar 2 × 2 matrix in the upper left corner.

$\begin{bmatrix} r & 0 & 0 \\ 0 & r & 0 \\ 0 & 0 & 1 \end{bmatrix}$	$\begin{bmatrix} \cos\theta & -\sin\theta & 0 \\ \sin\theta & \cos\theta & 0 \\ 0 & 0 & 1 \end{bmatrix}$	$\begin{bmatrix} 1 & 0 & 0 \\ 0 & -1 & 0 \\ 0 & 0 & 1 \end{bmatrix}$	$\begin{bmatrix} r & 0 & 0 \\ 0 & 1 & 0 \\ 0 & 0 & 1 \end{bmatrix}$
Dilation	Rotation Left	*X*-Axis Reflection	Horizontal Stretch

Add translation to this list, and now all of these simple transformations are 3 × 3 matrices. The advantage of using 3 × 3 matrices for all the transformations is that you can multiply the matrices to form a sequence of transformations.

Remember, matrix multiplication reads from right to left.

5. The Rotation/Dilation page of the sketch contains alternating red and blue matrices (**A** through **E**), each representing one step in a sequence of transformations. The black matrix **M** is their product (**M** = **EDCBA**). All of the matrices are initially set to the identity matrix.

The coordinates of independent points *P* and *Q* have already been measured, and point *P* has been transformed by matrix **M** to create *P′*. Because **M** is initially the identity matrix, *P* and *P′* are initially identical.

Be sure to record the formulas and not just numeric values.

Q2 In the next steps, you will edit the matrices to make point *P* rotate around *Q*. Record all of the new matrix definitions on paper.

6. Make matrix **A** a translation by vector *QO*. Notice the changes in **M** and in *P′*.

7. Make matrix **B** a rotation on the origin by angle θ. Observe the result when you drag the control for θ.

8. Make matrix **C** a translation by vector *OQ*. This is the inverse of matrix **A.**

Drag the control for angle θ. Because matrices **D** and **E** are still identities, you have effectively made matrix **M** equal to the product of the first three matrices (**M** = **CBA**).

Now that you have defined the matrix, you can apply it to any other point to perform the same transformation. You can use the custom tool **Transform Point** to do this without even measuring coordinates.

The segments give you a better look at the transformation.

9. Construct a polygon. Choose the **Transform Point** tool, and click each vertex. Connect the image points with segments. Drag θ to observe the rotation.

Q3 Change the matrix transformation into a dilation, centered on point Q and using scale factor r. You can make this change by editing matrix **B** only. What is the new definition for matrix **B**? Drag r to observe the dilation.

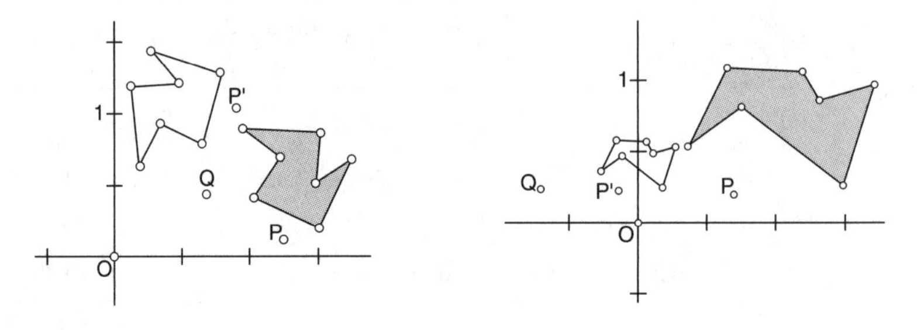

REFLECTION

10. On the Reflection page, points P, Q, and R have all been transformed by the matrix and their images have been plotted. (This will help you follow each step of the process.)

Q4 In the next steps, you will edit the matrices to create a reflection across line QR. Record all of the new matrix definitions on paper.

11. Make matrix **A** a translation by vector QO. Notice that the image of the line has moved along with the points.

This rotation to the right is equivalent to a left rotation of $-\theta$.

12. Make matrix **B** rotate line $Q'R'$ to the x-axis. Rather than measure any angles, work with the coordinates only. You will rotate to the right by angle θ. Use these formulas:

$$\sin\theta = \frac{y_R - y_Q}{\sqrt{(x_R - x_Q)^2 + (y_R - y_Q)^2}} \qquad \cos\theta = \frac{x_R - x_Q}{\sqrt{(x_R - x_Q)^2 + (y_R - y_Q)^2}}$$

13. Make matrix **C** a reflection across the x-axis.

14. Make matrix **D** the inverse of matrix **B**, and make matrix **E** the inverse of matrix **A**.

15. Construct a polygon. Choose **Transform Point.** Click on each vertex. Connect the image points with segments.

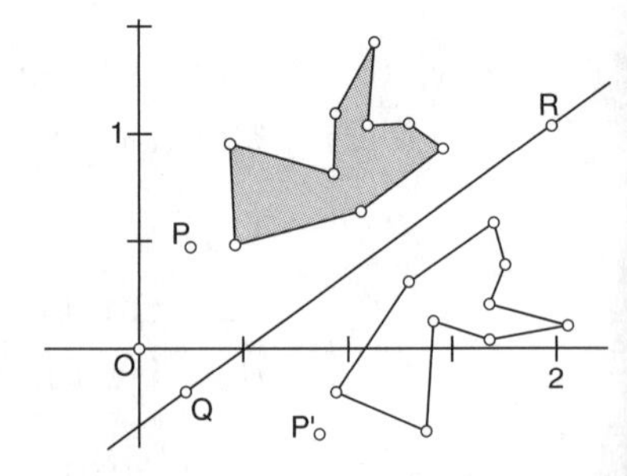

EXPLORE MORE

Page Blank has all of the matrices, but no points are defined yet. Create a transformation that will simulate a wheel rolling along the x-axis. Create other transformations of your own.

Coordinates in Three Dimensions

Just as you can represent points in a plane with rectangular or polar coordinates, you can represent points in three dimensions in several ways. Most common are rectangular (x, y, z), cylindrical (r, θ, z), and spherical (ρ, θ, ϕ) coordinates.

RECTANGULAR COORDINATES

1. Open the sketch **Coordinates 3D.gsp** in the **6 Vectors and Matrices** folder. This sketch contains quite a few objects, so spend a few minutes getting familiar with them.

The axes are displayed in perspective. You can use the action buttons to control the type of projection. You can use the three dials (*spin, pitch,* and *roll*) to control the viewing angle. Drag the sliders to control the viewing distance (in perspective mode), the display scale, and the length of the axes. When you first open the sketch, there are no projected objects other than the axes.

In three dimensions, rectangular coordinates are simply the planar (x, y) coordinates augmented by a third coordinate (z) representing height above the x-y plane.

2. Under the *Rectangular* heading are three parameters you can use as the rectangular coordinates of a point. Press the **Custom** tool icon at the bottom of the Toolbox, and choose the **Plot (x,y,z)** tool. Click on the coordinates, in order, to plot a point. Make the point red, matching the coordinates.

To show the Motion Controller, choose **Display | Motion Controller.**

3. Drag the viewing controls to change the viewing angle, and look at the coordinate system from different directions. Animate the spin control to put the coordinate system into motion, and use the Motion Controller to make the spinning slow.

Q1 You can change any of the three parameters defining the point by selecting the parameter and holding down the + or – key. How should the point move if you change the x-coordinate? What happens if you change the other coordinates? Test your predictions by modifying the coordinates.

CYLINDRICAL COORDINATES

Cylindrical (r, θ, z) coordinates are based on polar (r, θ) coordinates in a horizontal plane augmented by a z-coordinate representing height. One way to imagine plotting a point is to begin at the origin, move a distance r along the x-axis, rotate by angle θ to the left, and move a distance z upward.

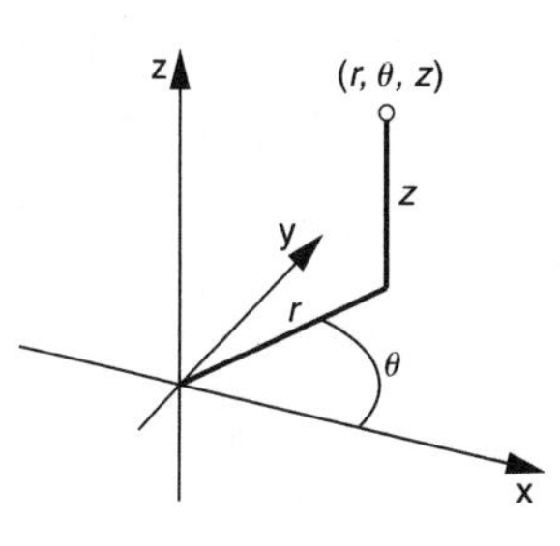

4. Below the *Cylindrical* heading are three parameters you can use as the cylindrical coordinates of a point. To plot the point, choose the custom tool **Plot (r,theta,z),** and click on the coordinates in order. Make the point blue, matching the coordinates.

To trace the point, select it and choose **Trace Point** from the Display menu.

Q2 How should the blue point behave if you gradually change one coordinate? Make a prediction, and test it for each coordinate. You may be able to see the changes more easily by tracing the point. Alternatively, you may prefer to have the coordinate system spinning slowly.

5. Set the (r, θ, z) coordinates to $(3.00, 1.40, -2.45)$. Now change the rectangular coordinates for the red point so that the points occupy the same position.

Q3 What are the (x, y, z) coordinates you used in the preceding step? Are the points really in the same position? Adjust the viewing controls to confirm this. The points should overlap no matter what angle you view them from.

You can rewrite the cylindrical coordinates in (x, y, z) form using these equations:

$$x = r\cos\theta \qquad y = r\sin\theta \qquad z = z$$

Notice that the z-coordinate is the same in both coordinate forms. These equations define a mapping from cylindrical coordinates to rectangular coordinates.

Q4 Did your answer to Q3 agree with the coordinates you calculate from these equations? If not, use the mapping equations to calculate rectangular coordinates corresponding to the cylindrical coordinates $(3.00, 1.40, -2.45)$. Check your answer in the sketch.

SPHERICAL COORDINATES

With spherical (ρ, θ, ϕ) coordinates, there is only one distance dimension, ρ (rho), measured from the origin to the point. The coordinate θ is the angle from the positive x-axis to the projection of the point onto the x-y plane. The coordinate ϕ (phi) is the angle from the positive z-axis to the point.

For any given point, θ is the same in cylindrical or spherical. However, don't confuse ρ with r. The spherical coordinate ρ is measured from the point to the origin, but the cylindrical coordinate r is measured from the point to the z-axis.

As with the cylindrical coordinates, there is a set of equations for mapping from spherical coordinates to rectangular coordinates:

$$x = \rho \sin\phi \cos\theta \qquad y = \rho \sin\phi \sin\theta \qquad z = \rho \cos\phi$$

6. Below the *Spherical* heading are three parameters you can use as the spherical coordinates of a point. To plot the point, choose the custom tool **Plot (rho,theta,phi),** and click on the coordinates in order. Color the point green, matching the coordinates.

Q5 What are your predictions for the movement of the green point as each of the (ρ, θ, ϕ) coordinates is changed? Test your predictions.

Q6 Given a point with spherical coordinates (3.77, 5.25, 2.40), what are the rectangular coordinates for the same point? Check your answer in the sketch.

EXPLORE MORE

Create a parameter *s* representing the edge length of a cube. Compute coordinates of the vertices using each of the three coordinate systems. Plot the coordinates, and construct the edges with the **Segment** tool.

Create coordinates representing some other shape, and use the Custom tools to plot them in three dimensions using the coordinate system of your choice. Animate the viewing controls to see the result.

Parametric Functions in Three Dimensions

The parametric variable is often called *t*. This activity uses θ because Sketchpad's Calculator can easily create functions of θ.

In this activity, you'll extend the concept of parametric functions to three dimensions. You'll plot parametric curves in rectangular (x, y, z) coordinates, using a single variable parameter θ. You'll use three functions of θ to determine the x, y, and z coordinates of the points on the curve.

As an additional exploration, you can create parametric curves using cylindrical (r, θ, z) and spherical (ρ, θ, ϕ) coordinates.

PARAMETRIC LINES

1. Open the Rectangular page of **Parametric 3D.gsp** in the **6 Vectors and Matrices** folder. Press the *Spin Axes* button to put the coordinate system in motion, and look at the other controls in the sketch.

Double-click the values or functions to edit them.

On the left of the screen are the three functions for x, y, and z. Below them are buttons to control the parameter, and values you can use to set its upper and lower limits. Four calculations show the current values of θ, $x(\theta)$, $y(\theta)$, and $z(\theta)$.

2. When you first open the sketch, you'll see these function definitions:

$$x(\theta) = \theta \qquad y(\theta) = \theta \qquad z(\theta) = \theta$$

Drag point θ, and observe how the graph of the parametric function appears, extending from the lower limit to the current value of θ. Press the *Advance* button to slowly increase θ, and the *Reset* button to return it to the lower limit.

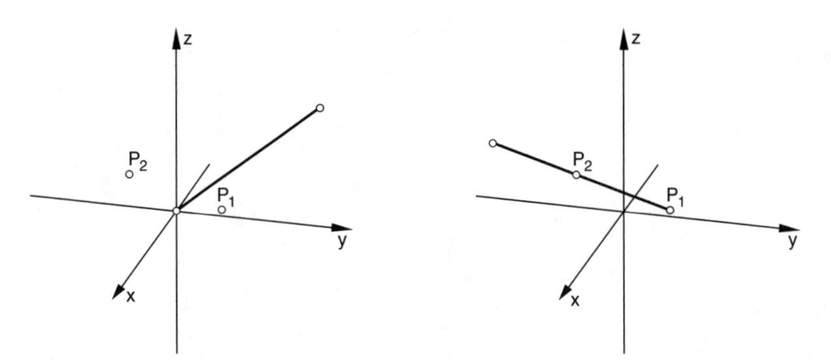

Q1 What shape do these parametric functions draw?

To enter a value like x_1 into the function definition, click the value in the sketch rather than entering a number. That way, the relation will hold even if the value changes.

3. Reset θ to its lower limit, click the *Show Points* button, and edit the function definitions as shown here:

$$x(\theta) = x_1 + (x_2 - x_1)\theta \qquad y(\theta) = y_1 + (y_2 - y_1)\theta \qquad z(\theta) = z_1 + (z_2 - z_1)\theta$$

Q2 Where will the blue point be when $\theta = 0$? Where will it be when $\theta = 1$? What figure do you expect to see when θ is animated? Test your prediction.

Q3 How can you change the limits of θ to make P_1 and P_2 endpoints of the path?

THE HELIX

4. Hide points P_1 and P_2 and their coordinates. Change the settings to the following:

upper limit = 20	$x(\theta) = \cos\theta$
lower limit = 0	$y(\theta) = \sin\theta$
	$z(\theta) = 0$

Q4 Were you surprised? The *x* and *y* functions describe a circle in the *x-y* plane. Because *z* is always zero, the curve stays in the *x-y* plane. What will happen if you force the curve to leave the *x-y* plane by making *z* a linear function of the variable?

5. Change the definition of *z* to $z(\theta) = \theta/5$.

This path has a corkscrew shape, called a *cylindrical helix*. You can inspect the function to see why. As the point is spinning circles in the *x-y* plane, it is slowly rising vertically, giving it the *helical* shape.

z
y
x

Q5 The curve in this sketch makes slightly more than three turns. How can you edit the function definitions (not the range), to make the curve complete exactly five turns?

Q6 This helix has a right-hand thread—it turns to the right as it goes downward, just as most screws do. How can you give it a left-hand thread?

Q7 How can you change the radius and height? Change the function definitions to create a left-handed helix that makes exactly five turns, with radius 3 and height 8.

THE CONE

Although the parametric curve is in three dimensions, it has no area or volume. A parametric function cannot describe a surface such as a cone, but it can bend a curve to fit onto a cone. You just need to adjust the curve so that its radius changes at a linear rate.

6. Enter these settings:

$$upper\ limit = 10$$

$$lower\ limit = 0$$

$$x(\theta) = \frac{\theta}{5}\cos 5\theta$$

$$y(\theta) = \frac{\theta}{5}\sin 5\theta$$

$$z(\theta) = \frac{\theta}{2}$$

Q8 How can you modify these settings to show both nappes of a double cone?

EXPLORE MORE

These images show parametrically defined curves winding around a sphere and a torus. Try to duplicate them.

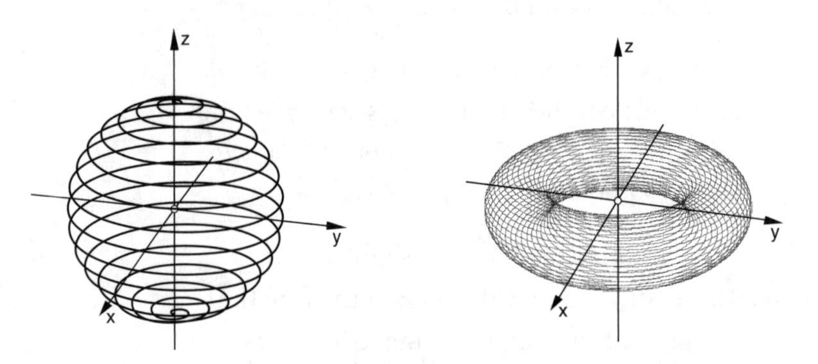

In the cylindrical and spherical sketches, the parametric variable is *x* in order to avoid confusion with the coordinate θ.

The other two pages of **Parametric 3D.gsp** have curves defined by parametric functions in cylindrical and spherical coordinates. Some of the curves shown here are actually easier to create in these alternate coordinate systems. Try to duplicate some of the sketches you've already done. Then use these alternate coordinate systems to create something new.

Vector Operations in Three Dimensions

These operations can also be done in higher dimensions, but are difficult to represent on the screen.

The vector operations you have used in two dimensions (addition, multiplication by a scalar, and the dot product) also work in three dimensions. An additional operation in three dimensions—the *cross product*—has no equivalent in two dimensions.

THE CROSS PRODUCT

1. Open **Vector Operations 3D.gsp** in the **6 Vectors and Matrices** folder.

The Cross Product page contains two vectors, **a** and **b.** Their initial points are at the origin, so the vector components are the same as the coordinates of their terminal points *A* and *B*. You can edit the components to change the vectors.

The next two steps construct the cross product **a** × **b.** Unlike the dot product (which is a scalar), the cross product is itself a vector, with components c_1, c_2, and c_3.

To label the calculations, select them and choose **Display | Label Measurements.** Type c[1] for the first label.

2. Press and hold the **Custom** tool icon and choose **3D Cross Product.** Click, in order, on the components of **a** and the components of **b.** Label the resulting calculations c_1, c_2, and c_3.

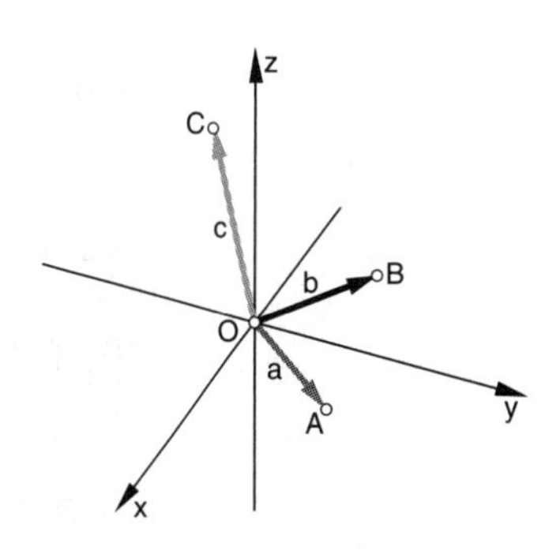

3. Choose the **Plot(x,y,z)** custom tool, and click in order on the components of **c.** Label the plotted point *C*. Construct segment *OC* and label it **c.** Use the **Arrowhead** custom tool to make the arrowhead.

4. Press the *Spin* button to put the axes in motion, and observe the relationships among **a, b,** and **c.**

Q1 Vector **c** should be perpendicular to both **a** and **b.** How can you use the dot product to confirm this? Use the **3D Dot Product** custom tool to compute **a** · **c** and **b** · **c.** What are the results?

Q2 Compute and plot **b** × **a.** What is its relationship to **a** × **b**?

To change a coordinate, double-click it, or select it and press the + or – key on the keyboard.

Q3 You can change vectors **a** and **b** by changing the coordinates of points *A* and *B*. Experiment with different nonzero coordinates to make **a** × **b** the zero vector. What can you conclude about $\angle AOB$ in this configuration?

COLLINEARITY

The answer to Q3 gives you a vector-based test for the collinearity of points.

5. Go to the Collinearity page, which has three plotted points, *P*, *Q*, and *R*. You can edit the coordinates to determine the positions of the points. The coordinates have been used to compute two vectors: **a** = $\overrightarrow{PQ}$ and **b** = $\overrightarrow{PR}$.

You cannot visually confirm collinearity in 3D from one perspective. Use the *Spin* button to put the axes in motion.

6. Compute **a** × **b.** Call it **c** and label the components (c_1, c_2, c_3).

Q4 Are the points collinear? How can you be sure? Change the coordinates of *P*, *Q*, and *R* until you have three points that are collinear. What are their coordinates? What is the cross product **a** × **b** when the points are collinear?

COPLANARITY

Any three points are coplanar, but what about four? Given four points in space, you can use vector operations to determine whether they lie in a common plane.

7. Go to the Coplanarity page, which has coordinates for four points: *P*, *Q*, *R*, and *S*.

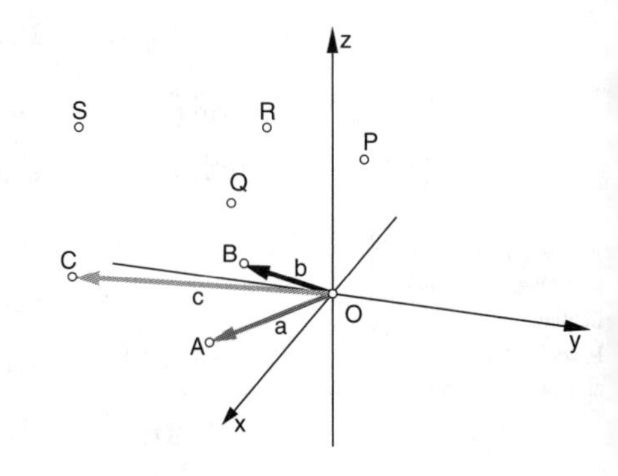

Step 7 effectively translates the given points by vector $\overrightarrow{PO}$.

8. Use the point coordinates to compute the components of three vectors: **a** = $\overrightarrow{PQ}$, **b** = $\overrightarrow{PR}$, and **c** = $\overrightarrow{PS}$. Use the **Plot(x,y,z)** custom tool to plot the components as terminal points *A*, *B*, and *C*. Construct the three vectors.

9. Compute **a** × **b,** and plot the components as point *N*. Draw the position vector and label it **n.**

Being perpendicular to both **a** and **b,** vector **n** must be normal to the plane defined by points *O*, *A*, and *B*. Any other vector in that plane must also be perpendicular to **n.**

Q5 Use the **3D Dot Product** custom tool to compute **c** · **n.** What is the value of this dot product when c is perpendicular to **n**?

10. You can change vector **c** by moving point *S*. Change the coordinates of point *S* to make **c** perpendicular to **n.**

Q6 If **n** is perpendicular to **a, b,** and **c,** what does that tell you about points *O*, *A*, *B*, and *C*? Hence, what do you know about points *P*, *Q*, *R*, and *S*? How can you confirm this visually?

EXPLORE MORE

Hint: Compute the length of a vector projection.

Q7 In the Coplanarity section, suppose that the four given points are not coplanar. Determine the coordinate distance from point *S* to the plane formed by the other three points.

Q8 Return to the Collinearity section. How far is point *R* from line *PQ*?

Matrix Transformations in Three Dimensions

You can use matrices to transform coordinates in three dimensions, just as you use them in the plane. In this activity you will use 3 × 3 matrices to transform the three-dimensional coordinates of points.

SIMPLE TRANSFORMATIONS

1. Open **Matrix Transformations 3D.gsp** in the **6 Vectors and Matrices** folder.

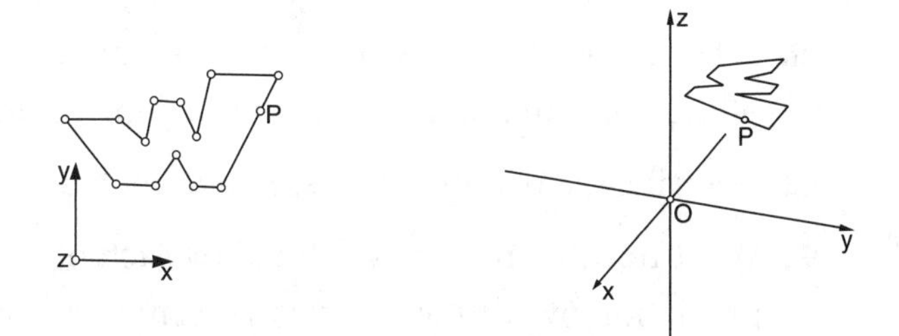

The sketch shows two separate views of a polygon. At the top is a two-dimensional view perpendicular to the *z*-axis. Below is a perspective view.

2. Drag point *P* in the top view and observe the behavior of point *P* in the perspective view.

3. Drag the polygon vertices in the top view and see how the perspective image responds. Then experiment with changing z_P, the *z*-coordinate of *P*.

4. The points labeled *spin, pitch,* and *roll* control the perspective view. Drag them so that the *z*-axis is pointing directly toward you, with the *x*-axis pointing right and the *y*-axis pointing up. When you have the controls adjusted this way, the perspective view will match the top view.

Q1 What positions of the controls result in the perspective view matching the two-dimensional view?

5. Use **Edit | Undo** to return the perspective controls to their original state.

You will now use the transformation matrix to transform point *P*.

6. Double-click the middle element of the matrix and change it to −1:

$$\begin{bmatrix} 1 & 0 & 0 \\ 0 & -1 & 0 \\ 0 & 0 & 1 \end{bmatrix} \begin{bmatrix} x_P \\ y_P \\ z_P \end{bmatrix}$$

Q2 What image do you think will result from this transformation?

To choose a custom tool, press and hold the custom tool button at the bottom of the toolbar.

7. Choose the custom tool **3D Matrix*Vector.** Click on the elements of the matrix, column by column, from top to bottom, and then on the coordinates of *P*. Three new coordinates appear. Label them x', y', and z'.

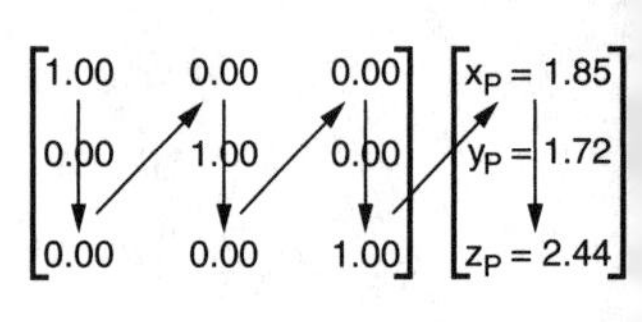

8. Choose the custom tool **Plot(x,y,z)** and click on the new coordinates in order. This will plot the point on the 3D axes.

9. Select *P* and the newly plotted point, and construct the locus. Drag the vertices of the polygon and adjust the view controls to help you visualize the result.

Q3 Describe the resulting transformation. Was it what you expected?

To insert a calculation in place of a matrix element, select the element and choose **Edit | Edit Parameter.**

Q4 What transformation is defined by each of the following matrices? Check each prediction by editing the matrix elements. For some matrices, you will need to replace some parameters with calculations using the values of angle θ or ratio r.

$\mathbf{A} = \begin{bmatrix} r & 0 & 0 \\ 0 & r & 0 \\ 0 & 0 & r \end{bmatrix}$	$\mathbf{B} = \begin{bmatrix} r & 0 & 0 \\ 0 & 1 & 0 \\ 0 & 0 & 1 \end{bmatrix}$	$\mathbf{C} = \begin{bmatrix} \cos\theta & -\sin\theta & 0 \\ \sin\theta & \cos\theta & 0 \\ 0 & 0 & 1 \end{bmatrix}$	$\mathbf{D} = \begin{bmatrix} -1 & 0 & 0 \\ 0 & 1 & 0 \\ 0 & 0 & -1 \end{bmatrix}$
$\mathbf{E} = \begin{bmatrix} \cos\theta & 0 & \sin\theta \\ 0 & 1 & 0 \\ -\sin\theta & 0 & \cos\theta \end{bmatrix}$	$\mathbf{F} = \begin{bmatrix} -1 & 0 & 0 \\ 0 & -1 & 0 \\ 0 & 0 & -1 \end{bmatrix}$	$\mathbf{G} = \begin{bmatrix} 1 & 0 & 0 \\ 0 & 1 & 0 \\ 0 & 0 & 0 \end{bmatrix}$	$\mathbf{H} = \begin{bmatrix} 1 & 0 & 0 \\ 0 & \cos\theta & -\sin\theta \\ 0 & \sin\theta & \cos\theta \end{bmatrix}$

COMBINED TRANSFORMATIONS

In this section you will dilate the polygon using an arbitrary center of dilation.

10. Go to the Combo page of **Matrix Transformations 3D.gsp.**

This page has an additional point *Q*, which has already been plotted in the perspective view. You will dilate point *P* by scale factor *r*, using *Q* as the center of dilation. To use *Q* as the center of dilation, you will first translate point *Q* to the origin, then dilate about the origin, and finally translate again to return point *Q* to its original position.

11. Subtract the coordinates of *Q* from the corresponding coordinates of *P*. Label the result (x_1, y_1, z_1). This is the translation of point *P* by vector $\overrightarrow{QO}$.

$$x_1 = x_P - x_Q$$

$$y_1 = y_P - y_Q$$

$$z_1 = z_P - z_Q$$

12. Verify that your calculated coordinates really represent the translation of point P by vector $\overrightarrow{QO}$. To do so, use the custom tool **Plot(x,y,z)** to plot the point P_1 defined by these coordinates, construct segments PP_1 and QO, and verify in the perspective view that these two vectors are always identical.

Q5 Describe the relationship between P_1 and O in terms of P and Q.

13. Once you've verified the translation, to avoid cluttering your sketch, delete the two segments.

14. Form the matrix for dilation by scale factor r, and multiply it by the vector (x_1, y_1, z_1).

$$\begin{bmatrix} r & 0 & 0 \\ 0 & r & 0 \\ 0 & 0 & r \end{bmatrix} \begin{bmatrix} x_1 \\ y_1 \\ z_1 \end{bmatrix} = \begin{bmatrix} x_2 \\ y_2 \\ z_2 \end{bmatrix}$$

15. Use the custom tool **Plot(x,y,z)** to plot these coordinates, and label the plotted point P_2.

Q6 Drag point r to change the scale factor, and describe the behavior of point P_2.

16. Add the coordinates of Q to translate by vector $\overrightarrow{OQ}$.

$$x_3 = x_2 + x_Q$$

$$y_3 = y_2 + y_Q$$

$$z_3 = z_2 + z_Q$$

17. Choose custom tool **Plot(x,y,z)** and plot (x_3, y_3, z_3). Label it P_3. Construct the locus of P_3 as P moves around the polygon.

Q7 What happens when you drag the r slider?

Q8 What will happen if you change the matrix to an x-axis rotation? Try it.

EXPLORE MORE

In the previous section you performed a sequence of three transformations: translate/dilate/translate, where the last translation is the inverse of the first. Open a new copy of the sketch and try this variation:

a. Rotate $(+45°)$ about the y-axis.

b. Rotate by angle θ about the x-axis.

c. Rotate $(-45°)$ about the y-axis.

Again, the first and last transformations are inverses of each other. You will have to create a separate matrix for each of the three transformations.

Matrix Products in Three Dimensions

You can do three-dimensional rotation, dilation, and reflection using 3×3 matrices, but translation requires 4×4 matrices.

In two dimensions, you can use multiplication of 3×3 matrices to perform general transformations, including translation, rotation, dilation, and reflection. In this activity you will use multiplication of 4×4 matrices to perform general transformations of three-dimensional objects.

THREE-DIMENSIONAL TRANSFORMATION MATRICES

$\begin{bmatrix} 1 & 0 & 0 & 0 \\ 0 & \cos\theta & -\sin\theta & 0 \\ 0 & \sin\theta & \cos\theta & 0 \\ 0 & 0 & 0 & 1 \end{bmatrix}$ Rotation about x-axis	$\begin{bmatrix} \cos\theta & 0 & \sin\theta & 0 \\ 0 & 1 & 0 & 0 \\ -\sin\theta & 0 & \cos\theta & 0 \\ 0 & 0 & 0 & 1 \end{bmatrix}$ Rotation about y-axis	$\begin{bmatrix} \cos\theta & -\sin\theta & 0 & 0 \\ \sin\theta & \cos\theta & 0 & 0 \\ 0 & 0 & 1 & 0 \\ 0 & 0 & 0 & 1 \end{bmatrix}$ Rotation about z-axis	$\begin{bmatrix} 1 & 0 & 0 & a \\ 0 & 1 & 0 & b \\ 0 & 0 & 1 & c \\ 0 & 0 & 0 & 1 \end{bmatrix}$ Translation by vector $[a, b, c]$
$\begin{bmatrix} r & 0 & 0 & 0 \\ 0 & r & 0 & 0 \\ 0 & 0 & r & 0 \\ 0 & 0 & 0 & 1 \end{bmatrix}$ Dilation by scale factor r	$\begin{bmatrix} -1 & 0 & 0 & 0 \\ 0 & 1 & 0 & 0 \\ 0 & 0 & 1 & 0 \\ 0 & 0 & 0 & 1 \end{bmatrix}$ Reflection across y-z plane	$\begin{bmatrix} 1 & 0 & 0 & 0 \\ 0 & -1 & 0 & 0 \\ 0 & 0 & 1 & 0 \\ 0 & 0 & 0 & 1 \end{bmatrix}$ Reflection across x-z plane	$\begin{bmatrix} 1 & 0 & 0 & 0 \\ 0 & 1 & 0 & 0 \\ 0 & 0 & -1 & 0 \\ 0 & 0 & 0 & 1 \end{bmatrix}$ Reflection across x-y plane

1. Open **Matrix Product 3D.gsp** in the **6 Vectors and Matrices** folder.

Remember, matrix multiplication works from right to left.

The sketch contains five matrices and their product (**M** = **EDCBA**). When you open the sketch, all of the matrices are identity matrices, and so is their product.

The sketch contains a custom tool that uses three coordinates (x, y, and z) and transforms and then plots the point defined by the coordinates.

PLANETARY ORBIT

2. Create parameters with values -1, 0, and 1.

3. Choose the **Transform & Plot (x,y,z)** custom tool. Plot the x-axis unit point $(1, 0, 0)$ by clicking in order on the parameters 1, 0, and 0. (To click on the 0 parameter twice in a row, click it once, move the mouse, and then click it again.) Use this tool to plot the unit points, positive and negative, on all three axes:

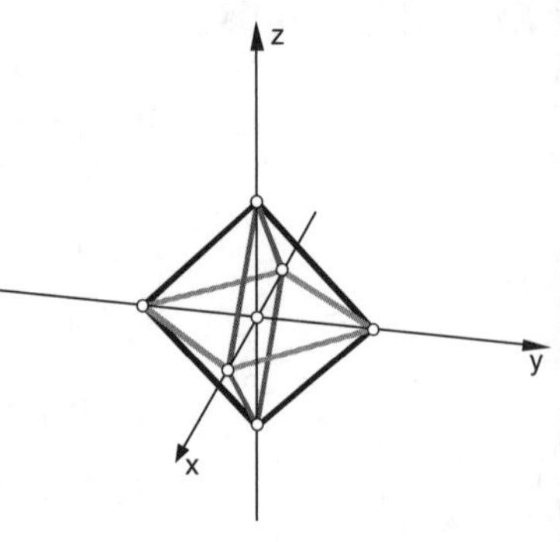

$$(1, 0, 0)\ (0, 1, 0)\ (0, 0, 1)\ (-1, 0, 0)\ (0, -1, 0)\ (0, 0, -1)$$

4. These six points are vertices of a regular octahedron. Construct the edges with line segments.

In the next steps, you will apply several transformations to make the octahedron mimic the motion of a planet. The planet will rotate on its axis ten times in each orbit.

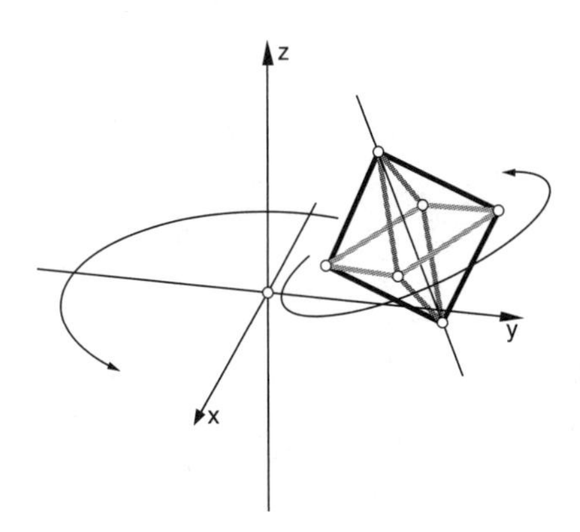

You can adjust angle θ and scalar r with the circle and slider controls below the matrices.

5. Edit matrix **A** to create a rotation by angle 10θ about the z-axis. Drag the angle control, and observe its effect.

6. Make matrix **B** a rotation by 23° about the x-axis. This gives the axis a tilt approximately the same as Earth's.

7. Make matrix **C** a rotation by angle $-\theta$ about the z-axis. Drag the angle control again. This causes the axis itself to change direction. The reason for this will become clear after the last transformation.

8. Make matrix **D** a translation by vector $(r, 0, 0)$.

9. Make matrix **E** a rotation by angle θ about the z-axis. This causes the planet to orbit, and it also undoes the rotation in matrix **C.**

10. Animate the angle control. You can change the look by adjusting the sliders controlling the scale and the measurement r.

EXPLORE MORE

When you use the custom tool **Transform & Plot (x,y,z),** you can apply the transformation that you defined in this activity to any new points. Use the tool to construct a moon for the planet. Make it in the shape of some other polyhedron, maybe a cube or a tetrahedron.

Page 2 of **Matrix Product 3D.gsp** is the same as the first page except that it uses the product of ten matrices (**M** = **KJHGFEDCBA**). Use it to define more complex transformations. Here are some suggestions:

Make a polyhedron wind around the surface of a torus while the torus itself is rotating and orbiting the origin.

Create a rotation about an axis defined by two arbitrary points.

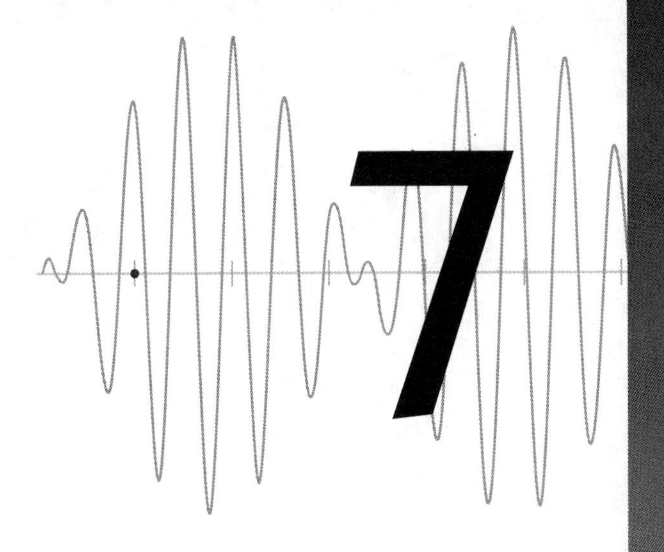

7

Polar Coordinates and Complex Numbers

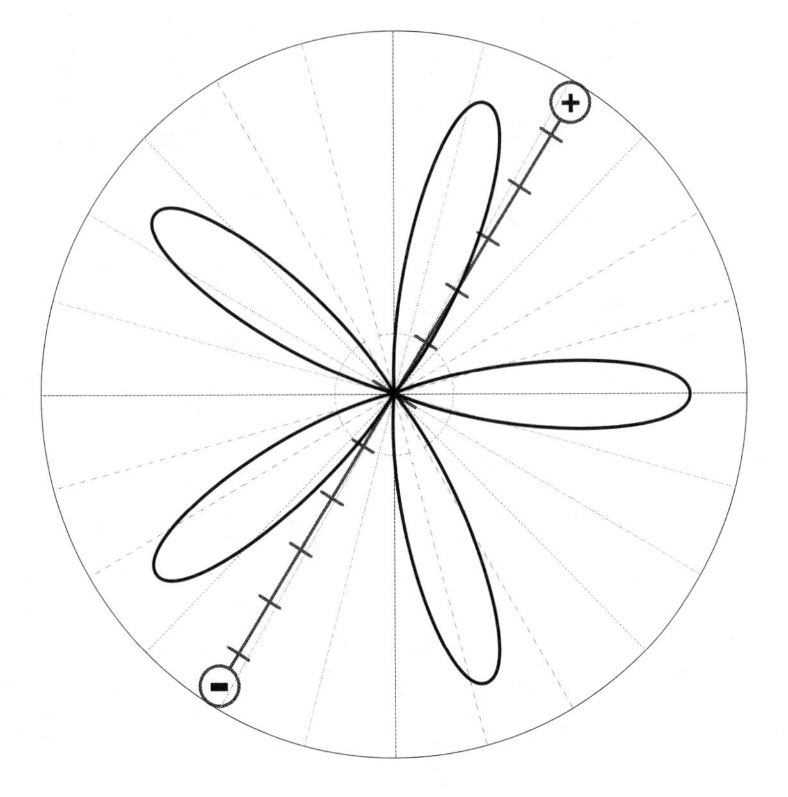

Introduction to Polar Coordinates

In one dimension—on a line—you need only one number to specify a location. For instance, point *A* at the right is located at 1.4. You can specify any point on the number line this way. In two dimensions, you need two numbers to specify a location. You're familiar with one way of doing this: using *rectangular* (x, y) coordinates. In this activity, you'll learn about another way: using *polar* (r, θ) coordinates.

is a Greek letter, pelled "theta" n English.

SKETCH AND INVESTIGATE

1. In a new sketch, choose **Edit | Preferences,** and change the settings to match those at right.

2. Create a polar grid by choosing **Graph | Grid Form | Polar Grid.**

se the **Graph | New arameter** command. Make sure to set the nits to degrees when ou create the *theta* arameter.

3. Create two new parameters, one labeled *r* with a value of 4 and one labeled *theta* with a value of 45°.

Q1 In the next step, you'll plot the point with polar coordinates (4, 45°). Before proceeding, make a conjecture as to where that point will be located.

4. Select, in order, *r* and *theta.* Choose **Graph | Plot As (r, theta).** Was your conjecture correct? Label the new point *A*, and connect the *pole* (the origin) with point *A* using a thick segment.

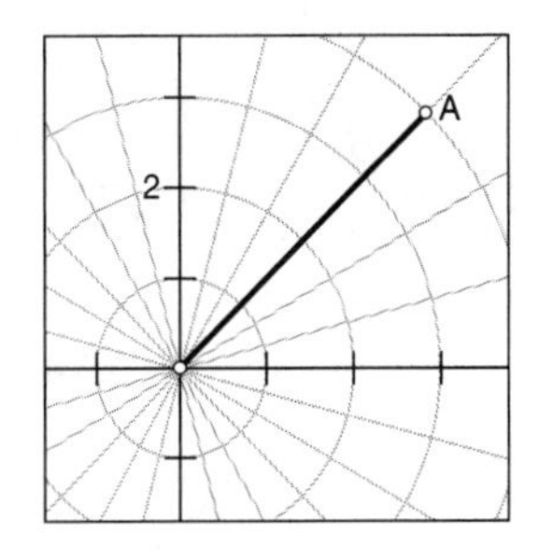

5. Explore parameter *r* by selecting it and pressing the **+** and **–** keys on your keyboard. Explore both positive and negative values of *r.* Explore *theta* similarly.

Q2 Describe how both *r* and θ work when plotting (r, θ) in polar coordinates.

6. Plot the point (3, 150°) using the **Graph | Plot Points** command. Label this point *target.*

ou can double-click a arameter to change it.

Q3 Set parameter *r* to 3. Can you get point *A* to land on point target by making parameter *theta* something other than 150°? If so, list coordinates in the form $(3, \theta)$ for three such points.

Q4 Set parameter *theta* to 150°. Can you get point *A* to land on point *target* by making parameter *r* something other than 3? If so, list coordinates in the form $(r, 150°)$ for three such points.

Q5 Can you get *A* to land on target with $r \neq 3$ and *theta* $\neq 150°$? If so, list coordinates in the form (r, θ) for three such points.

CONVERT BETWEEN THE TWO FORMS

Now that you are more familiar with how polar coordinates work, you will develop formulas for converting between polar and rectangular coordinates.

7. Select point *A*, and measure its abscissa (*x*-coordinate) by choosing **Measure | Abscissa (x)**. Similarly, measure the ordinate (*y*-coordinate) of point *A*.

Choose **Graph | Calculate** to open the Calculator. Click on a measurement or parameter in the sketch to enter it into a calculation.

Q6 Using the drawing at right as a guide, write an expression for x in terms of r and θ. To check your answer, use the Calculator to calculate your expression with the *r* and *theta* parameters as inputs. The resulting calculation should always equal the x_A measurement from step 7, regardless of *A*'s location.

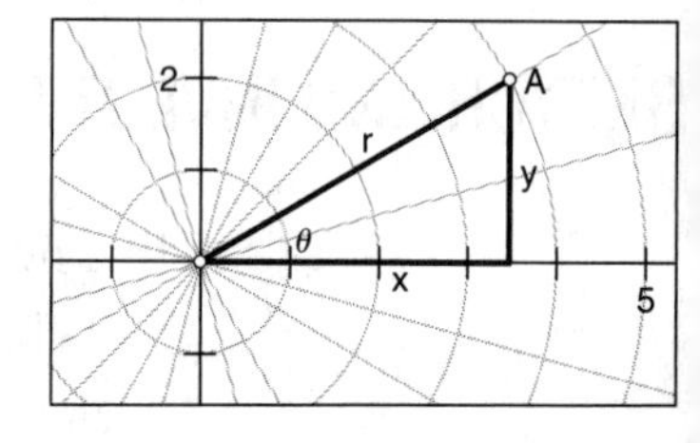

Q7 Write an expression for y in terms of r and θ. Use the Calculator to verify your expression, as in Q6.

Q8 Write an expression for r in terms of x and y. Use the Calculator to verify your expression, using the x_A and y_A measurements as inputs.

Q9 Write an expression for θ in terms of x and y. Use the Calculator to verify your expression, as in Q8.

EXPLORE MORE

Q10 In the Cartesian (x, y) plane, the graph of $x = 6$ is a vertical line with an *x*-intercept of 6 because it is the set of all points whose *x*-coordinates equal 6 and whose *y*-coordinates take on any value. What will the graph of $r = 6$ look like in the polar plane? Make a guess, and test it using the **Graph | Plot New Function** command. Are there any other functions that will give the same graph?

To plot $\theta = 75°$, choose **θ = f(r)** from the Equation pop-up menu of the New Function dialog box. Once plotted, select the plot and choose **Edit | Properties.** On the Plot panel, change the domain so that *r* takes on both negative and positive values.

Q11 What do you think the graph of $\theta = 75°$ will look like? Again, guess first and then use the **Graph | Plot New Function** command to test your answer. Are there any other functions that will give the same graph?

Q12 Given what you learned in Q6 and Q7, can you graph the equivalent of $x = 6$ using polar coordinates? How about the equivalent of $y = -2$?

Q13 What function do you think gives the graph at right? Make a guess, and test it using the **Graph | Plot New Function** command. (*Hint:* What are the coordinates of the *x*- and *y*-intercepts?)

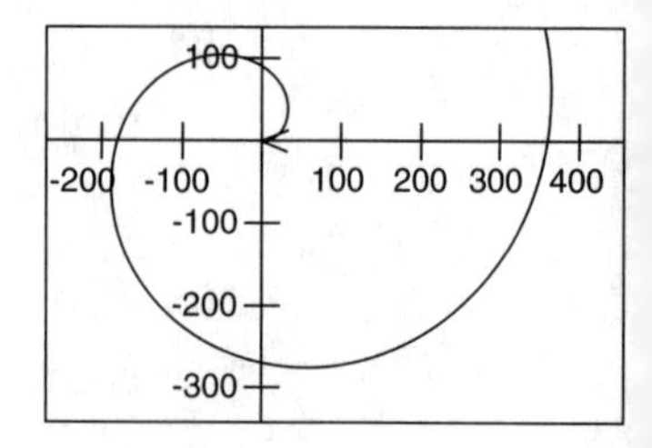

Cartesian Graphs and Polar Graphs

artesian coordinates are lso called *rectangular* oordinates.

You have probably done lots of graphing in Cartesian (x-y) coordinates, but relatively little in polar (r, θ) coordinates. In this activity you will explore various functions plotted in both coordinate systems. You will find connections between the two types of graphing and use your understanding of Cartesian graphing to better understand polar graphing.

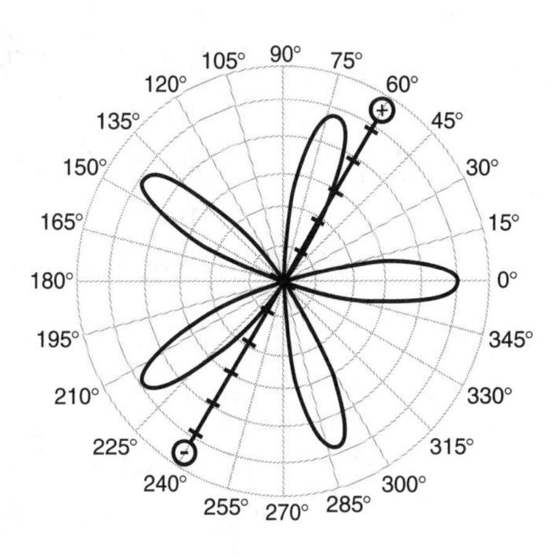

SKETCH AND INVESTIGATE

Start by investigating functions of the form $f(t) = a\cos(bt)$. You can think of these as $y = a\cos(bx)$ in the Cartesian plane and $r = a\cos(b\theta)$ in the polar plane.

Q1 Start with $a = 5$ and $b = 3$. Use what you know about graphing in the Cartesian plane to make an approximate sketch of $y = 5\cos(3x)$ on the following grid.

What do the parameters a and b control?

1. Open the sketch **Cartesian Polar.gsp** from the **7 Polar Coords and Complex Nos** folder. This sketch contains the plot of the function $y = a \cdot \cos(b \cdot x)$. When you open the sketch, parameters a and b are both set to 1, so the function being plotted is simply $y = \cos x$.

o adjust a parameter, ouble-click it, or select and press the + or – ey on your keyboard.

2. Adjust parameters a and b to check your plot from Q1. Was your sketch accurate?

3. Drag the purple pointer back and forth along the number line. Note how the red bar shows the input (x) value and the green "bowtie" shows the output (y) value of the function. The **+** and **–** signs on the red bar show where y is positive and negative, respectively. Move the pointer back to 0°.

Q2 In a moment, you'll reveal the corresponding polar curve, $r = 5\cos(3\theta)$. First, make a wild guess about what it will look like, and write down your guess.

4. After you have made a guess, press the *Show Polar* button. Compare the result with your guess.

5. Drag the purple pointer back and forth slowly, this time looking for connections between the Cartesian and polar graphs. Note that the "bowtie" is always in the same relative position on the two output axes, but the red bar corresponding to x in the Cartesian plane slides right and left, and the red bar corresponding to θ in the polar plane spins around.

Q3 Which points on the polar graph correspond to x-intercepts on the Cartesian graph?

Q4 Which points on the polar graph correspond to maximum points on the Cartesian graph? Which points on the polar graph correspond to minimum points on the Cartesian graph? Is there any connection between these points?

Q5 Dragging the purple pointer from 0° to 360°, the polar graph repeats itself, with an interesting twist. What is different about the second copy?

Q6 Make a prediction as to how changing parameter a will affect the polar graph. Now adjust parameter a in the sketch, and see if you're right. What does the a-value appear to control in the polar sketch?

Don't be surprised if the answer to Q7 is a little more complicated than you may have expected!

Q7 Make a prediction as to how changing parameter b will affect the polar graph. Adjust parameter b in the sketch, and see if you're right. What does the b-value appear to control in the polar sketch? Be sure to try both odd and even values.

EXPLORE MORE

Q8 Explain the pattern you discovered in Q7. (*Hint:* Your answer to Q5 is relevant here. Think about what has to be true in order for the polar graph to repeat itself at 180° and for which values of b this can happen.)

Q9 Go to page 2 of **Cartesian Polar.gsp.** There you'll see $f(x) = \frac{2}{\cos(x)}$, which you could also write as $f(x) = 2\sec(x)$. Predict what the polar graph will look like. After you've recorded your prediction, press the *Show Polar* button to reveal the answer. Can you explain analytically why the graph looks this way?

Q10 Go to page 3 of **Cartesian Polar.gsp.** On this page you can enter your own functions to try them out. You can use the values of parameters a and b in these functions if you like. Try several different kinds of functions, and record your results.

Multiplication of Complex Numbers

By using the distributive property of multiplication and the fact that $i^2 = -1$, you can use algebra to compute the product of two complex numbers, $a + bi$ and $c + di$:

$$\begin{aligned}(a + bi)(c + di) &= ac + (bc + ad)i + bdi^2 \\ &= (ac - bd) + (bc + ad)i\end{aligned}$$

Algebra alone doesn't tell the whole story. In this activity, you'll explore complex number multiplication from a geometric perspective.

SKETCH AND INVESTIGATE

1. Open the sketch **Complex Multiplication.gsp** in the **7 Polar Coords and Complex Nos** folder. You'll see the complex number **w** $= a + bi$ and another complex number, **z** on the complex plane. Because complex numbers have both a length and a direction with respect to the origin, **w** and **z** are represented as vectors. Your goal is to construct the product, **zw.**

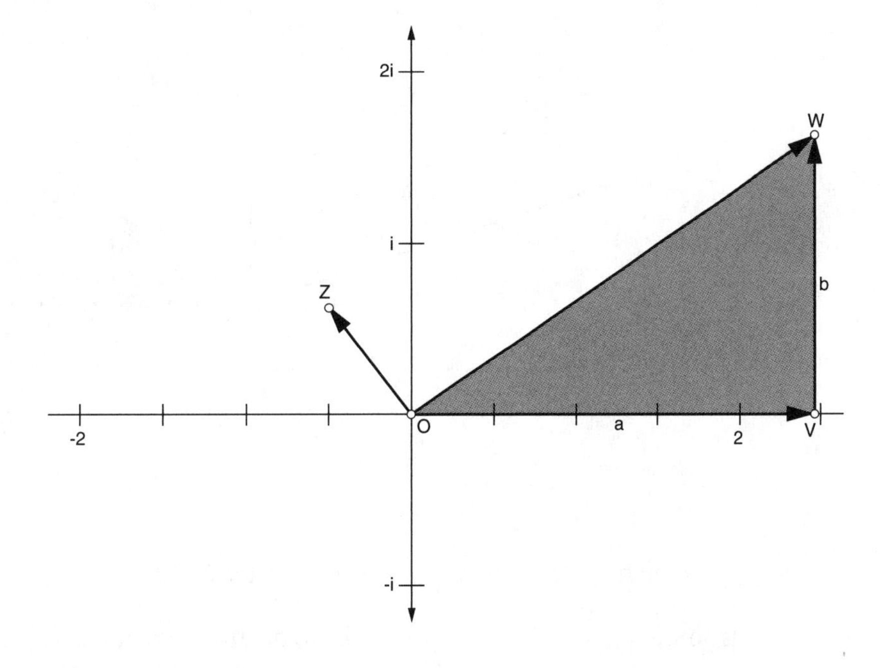

Algebraically, **zw** $=$ **z**$(a + bi) =$ **z**$a +$ **z**bi. In finer detail, multiplying **w** by **z** consists of four tasks:

Task 1: Multiply vector **z** by the scalar value a.

Task 2: Multiply vector **z** by i.

Task 3: Multiply vector **z**i by the scalar value b.

Task 4: Add vector **z**a and vector **z**bi.

The sketch comes with three pre-built tools for carrying out these steps visually on the complex plane. The tools, available by clicking on the **Custom** tools icon in the Toolbox, are listed here:

- **Scale:** Multiplies a vector (complex number) by a scalar value. To use the tool, click on the head of a vector and then the desired scalar value.
- **Multiply by *i*:** Multiplies a vector (complex number) by *i*. To use the tool, click on the head of the vector.
- **Add complex #s:** Adds together two vectors (complex numbers) and displays the result as a vector. To use the tool, click on the heads of the vectors.

2. Use the custom tools to multiply **w** by **z.** Follow, in order, the four tasks listed earlier. Your completed construction should look similar to the following picture:

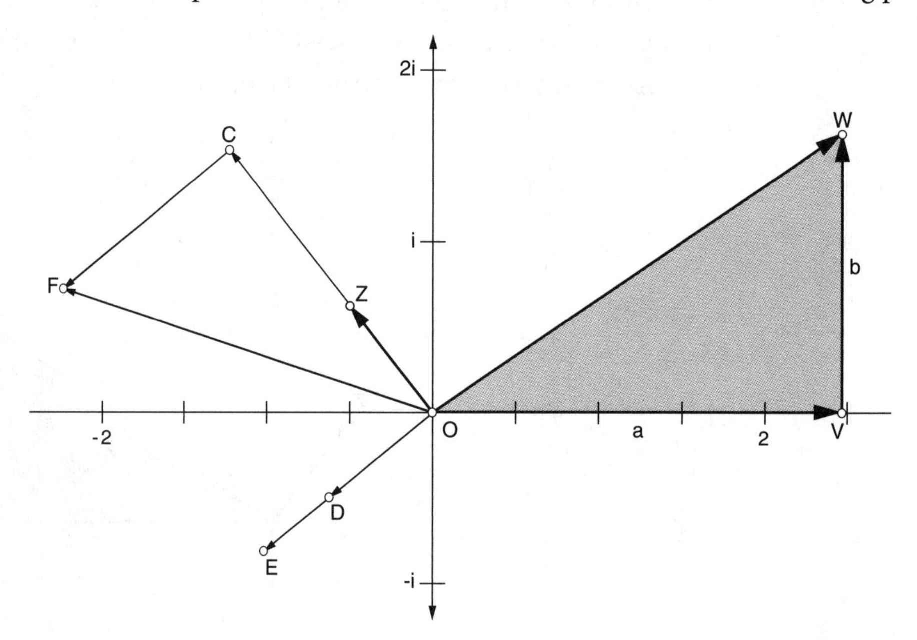

Q1 Which vector represents **wz** in the picture?

Triangle *OVW* has sides of lengths *a*, *b*, and $|\mathbf{w}|$, and *OZ* has length $|\mathbf{z}|$. Your answers to the next three questions should all be given in terms of these values.

Remember, multiplication by *i* is equivalent to a 90° rotation.

Q2 What is the length of $\overrightarrow{OC}$?

Q3 What are the lengths of $\overrightarrow{OD}$ and $\overrightarrow{OE}$?

Q4 What is the length of $\overrightarrow{CF}$?

Q5 What is the measurement of $\angle OCF$?

Q6 Explain why $\triangle OVW$ is similar to $\triangle OCF$.

Q7 Given that the two triangles are similar, what is the length of $\overrightarrow{OF}$?

Look back at your similar triangles.

Q8 The argument of a vector is the angle it makes with the positive real axis. If the argument of $\overrightarrow{OW}$ is α and the argument of $\overrightarrow{OZ}$ is β, then what is the argument of $\overrightarrow{OF}$?

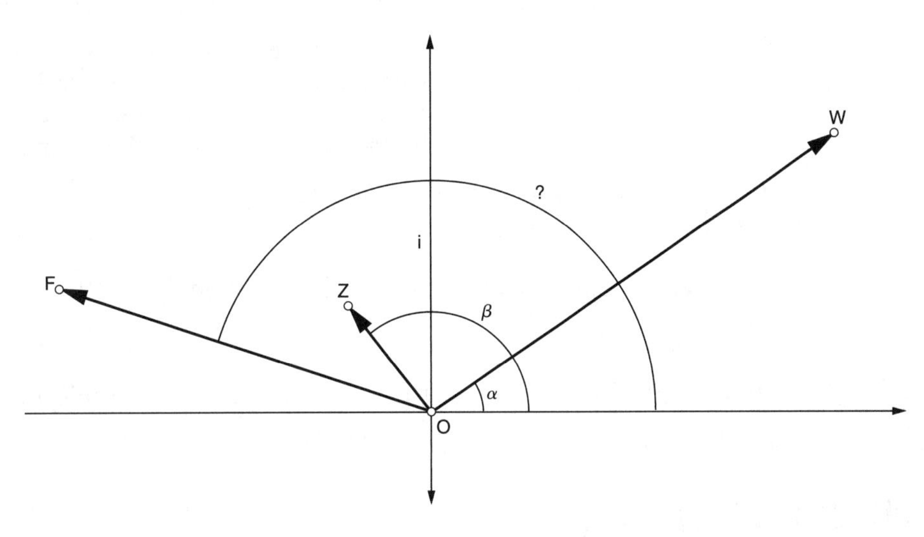

Q9 Complete this sentence:

To multiply two complex numbers **w** and **z**, __________ their lengths and __________ their arguments.

Q10 If $|\mathbf{w}| = 3$, $|\mathbf{z}| = 2$, $\alpha = 40°$, and $\beta = 110°$, then find the value of **wz**. Write your answer in the form $x + iy$.

EXPLORE MORE

Now that your sketch is complete, you can use it to explore the product of any two complex numbers.

3. Hide all vectors in your picture, leaving only $\overrightarrow{OZ}$, $\overrightarrow{OW}$, and $\overrightarrow{OF}$.

Q11 Find a complex number other than i whose square is equal to -1.

Q12 Find two different complex numbers a and b such that $a^2 = b^2 = i$.

Q13 Explain geometrically why the product of a complex number $a + bi$ and its reciprocal $a - bi$ is a real number.

In Search of Buried Treasure

Tucked away in the attic of your grandparents' farmhouse is a dusty old letter describing the location of a buried treasure. It reads:

> To find the treasure, walk into the field behind the farmhouse. Start at the scarecrow. Walk straight from the scarecrow to the oak tree. Count your steps as you walk. When you reach the oak tree, turn 90 degrees to the right and count off the same number of steps. When you reach the end, place a marker in the ground.
>
> Now return to the scarecrow. Walk straight from the scarecrow to the elm tree. Again, count your steps as you walk. When you reach the elm tree, turn 90 degrees to the left and count off the same number of steps. When you reach the end, place a marker in the ground.
>
> Connect the two markers with a rope. Dig beneath the midpoint of the rope to find the treasure.

The oak tree and the elm tree are still in the field, but the scarecrow is long gone. Is the treasure lost?

In this activity, you'll unearth the buried treasure by building a Sketchpad model and then prove your results by using complex numbers.

SKETCH AND INVESTIGATE

1. Open a new sketch and draw three random points to represent the scarecrow (point S), the oak tree (point O), and the elm tree (point E).

2. Draw segments OS and OE.

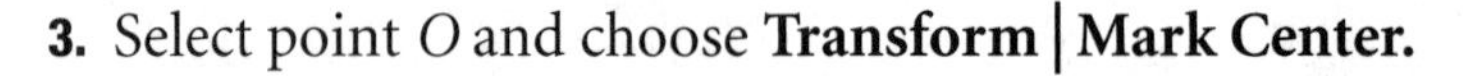

3. Select point O and choose **Transform | Mark Center.**

Choose **Rotate** from the **Transform** menu.

4. Select segment OS (including point S) and rotate it 90° about point O. Label the marker (the rotated point S) M_1.

To add a subscript at the end of a label, type the subscript enclosed in square brackets—for instance, M[1].

5. Rotate segment ES by $-90°$ about point E. Label the second marker M_2.

6. Construct point T—the location of the treasure—the midpoint of segment M_1M_2.

Q1 Drag point S around the screen. What do you notice about the treasure, point T?

PROVE YOUR RESULTS

7. Open the sketch **Buried Treasure.gsp** from the **7 Polar Coords and Complex Nos** folder. You'll see an arrangement similar to your own sketch, except in this sketch, the points and segments have been placed on the complex plane.

The axes have been arranged so that the oak tree and the elm tree sit at $(-1, 0)$ and $(1, 0)$, respectively.

Q2 Drag point *S* and observe whether point *T* moves. What is the location of point *T* on the complex plane?

Point *S* is free to move anywhere on the complex plane. We'll describe its location as (a, b) or, equivalently, $a + bi$.

8. Press the *Move Origin to Oak Tree* button, and watch the origin reposition itself 1 unit to the left.

Q3 Prior to moving the origin, the location of the scarecrow was $a + bi$. What is its new location?

Q4 By construction, M_1 is a rotation of point *S* by 90° about *O*. In the complex plane, multiplication by what number corresponds to a rotation of 90°?

Q5 Use your answer to Q4 to write the location of point M_1.

9. Press the *Move Origin Back* button, and watch the origin move 1 unit to the right, back to its original location.

Q6 What is the new location of point M_1?

10. Press the *Move Origin to Elm Tree* button, and watch the origin reposition itself 1 unit to the right.

Q7 Prior to moving the origin, the location of the scarecrow was $a + bi$. What is its new location?

Q8 By construction, point M_2 is a rotation of point *S* by $-90°$ about point *E*. In the complex plane, multiplication by what number corresponds to a rotation of $-90°$?

Q9 Based on your answer to Q8, what is the location of point M_2?

11. Press the *Move Origin Back* button, and watch the origin move 1 unit to the left, back to its original location.

Q10 What is the new location of point M_2?

Q11 Based on your answers to Q6 and Q10, compute the location of the treasure. Does your result match what you see in the sketch?

emember, *T* is the nidpoint of segment M_1M_2.

Transformations in the Complex Plane

In this activity, you'll explore the geometric effects of performing a variety of transformations on a complex number.

SKETCH AND INVESTIGATE

1. Open the sketch **Complex Transformations.gsp** in the **7 Polar Coords and Complex Nos** folder. The axes represent the complex plane with real numbers on the horizontal axis and complex numbers on the vertical axis.

Each of the ten pages shows a complex number $z = a + bi$ and another complex number w, which represents some transformation of z.

Q1. Drag z around the screen, and pay attention to the behavior of w. Match the ten different transformations of z to the following list:

a. $w = 2z$

b. $w = -z$

c. $w = a$ (the real component of z)

d. $w = bi$ (the complex component of z)

e. $w = |z|$ (the length of z)

f. $w = a - bi$ (the conjugate of z)

g. $w = z - 3$

h. $w = iz$

i. $w = z^2$

j. $w = \frac{1}{z}$

2. Open the sketch **Complex Transformations 2.gsp.** The sketch consists of six pages, each showing a transformation from Q1.

Q2. The complex number z is confined to the border of the letter *F*. Before you drag z, make a prediction: What will the corresponding path of w look like?

3. Select point w and choose **Display | Trace Point.** Drag point z, and observe the trace of point w.

4. Turn off tracing for point w. Then select points z and w, and choose **Construct | Locus.** Drag any vertex of the letter, and observe the effect on the locus of w.

Powers of Complex Numbers

If $z^2 = 1$, it's not hard to name two values of z that satisfy the equality. Both $z = 1$ and $z = -1$ work.

Now consider $z^3 = -1$. One solution is $z = -1$. There are no other real solutions. Are there other solutions in the complex numbers?

In this activity, you'll find solutions to equations like $z^3 = -1$ by exploring the geometric effects of raising complex numbers to integer powers.

PRELIMINARY WORK

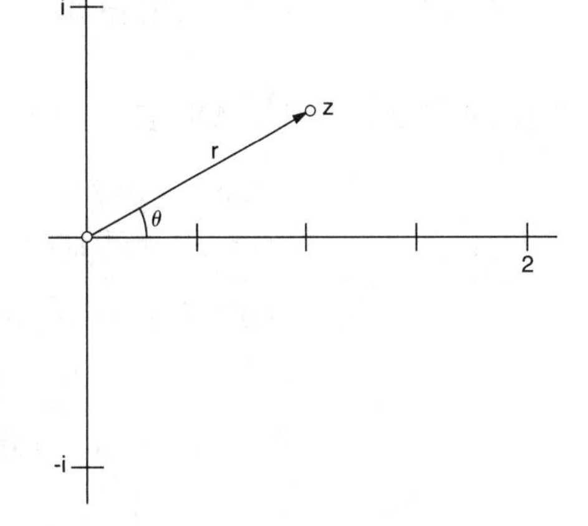

There are two common ways to represent a complex number z. The first is $z = a + bi$, where a and b are real numbers.

This second form is often written using the shorthand notation $re^{i\theta}$.

Another form is $z = r(\cos\theta + i\sin\theta)$. In this representation, z is a vector with length r that makes an angle of θ with the positive real axis. The angle θ is called the *argument* of z.

Q1 Complete this statement:

To multiply two complex numbers, __________ their lengths and __________ their arguments.

Q2 Complete this statement based on Q1:

To square a complex number z, __________ its length and __________ its argument.

Q3 Write $2i$ in the form $r(\cos\theta + i\sin\theta)$.

Q4 Use the information from Q2 and Q3 to find a complex number z that equals $2i$ when squared.

INVESTIGATE z^2

1. Open the sketch **Complex Powers.gsp** from the **7 Polar Coordinates and Complex Numbers** folder. The first page of the sketch shows the complex plane with real numbers on the horizontal axis and imaginary numbers on the vertical axis. The sketch contains two complex numbers, z and z^2.

Position point z as best you can so that $z^2 = 2i$. Use the arrow keys for small adjustments. Use the movement buttons for even smaller adjustments.

Q5 Drag point z to determine whether the value of z from Q4 does indeed satisfy $z^2 = 2i$. What is the value of z in the form $a + bi$?

If you need a hint, drag z to a different position for which $z^2 = 2i$.

Q6 Real numbers have both a positive and a negative square root. Complex numbers are neither positive nor negative, but a similar idea applies. Can you name another value of z that satisfies $z^2 = 2i$? Write your answer in the form $a + bi$.

Q7 Use your sketch to determine whether the value of z from Q6 satisfies $z^2 = 2i$. What is the value of z in the form $r(\cos\theta + i\sin\theta)$?

Use the method from Q2.

Q8 Check that the $r(\cos\theta + i\sin\theta)$ value of z from Q7 is indeed equal to $2i$ when squared.

2. Position the two red x's to indicate the two locations of z for which $z^2 = 2i$.

HIGHER POWERS OF *z*

Open page 2 of the sketch **Complex Powers.gsp.** You'll see three complex numbers represented as vectors: z, z^2 (not labeled), and z^3.

Q9 Drag point z to find three values of z for which $z^3 = 2i$. Write each in the form $a + bi$ and $r(\cos\theta + i\sin\theta)$.

Q10 Complete this statement:

To cube a complex number z, __________ its length and __________ its argument.

Use the method from Q10.

Q11 Look at each of the values from Q9 that you wrote in the form $r(\cos\theta + i\sin\theta)$. Cube each to check that $z^3 = 2i$.

3. Position the three red x's to indicate the three locations of z for which $z^3 = 2i$.

Q12 Open page 3 of the sketch. Find and verify the four values of z for which $z^4 = 2i$. Place red x's at these locations.

Q13 Open page 4 of the sketch. Find and verify the five values of z for which $z^5 = 2i$. Place red x's at these locations.

Q14 On all four pages of your sketch, connect consecutive red x's (the solutions to $z^n = 2i$) with segments, proceeding either clockwise or counterclockwise. Describe the shapes formed by the segments.

EXPLORE MORE

Q15 Use the sketch to solve $z^2 = -1$, $z^3 = -1$, $z^4 = -1$, and $z^5 = -1$.

A Geometric Approach to $e^{i\pi}$

One of the most beautiful—and mysterious—mathematical discoveries has to do with the value of $e^{i\pi}$. What does it mean to raise the number e to an imaginary power?

In this activity, you'll explore $e^{i\pi}$ through a geometric approach.

GETTING STARTED

A good place to begin your investigation is with the mathematical constant e. Consider these calculations:

$$\left(1+\frac{1}{10}\right)^{10}, \left(1+\frac{1}{100}\right)^{100}, \left(1+\frac{1}{1{,}000}\right)^{1{,}000}, \left(1+\frac{1}{10{,}000}\right)^{10{,}000}, \ldots$$

1. Open a new sketch. Choose **Edit | Preferences** from the Edit menu. Set the Scalars precision to **hundred-thousandths.**

2. Choose **Measure | Calculate.** Calculate, one at a time, the values of the four expressions above.

Q1 What do you notice about your four calculations?

3. Calculate four more expressions that continue the sequence above.

Q2 Based on your calculations, approximate the value of $\left(1+\frac{1}{n}\right)^n$ to several decimal places as n grows ever larger.

The mathematical constant e is defined as the limiting value of $\left(1+\frac{1}{n}\right)^n$ as n approaches infinity. Raising e to a power, like e^2 or e^3, involves a similar definition:

$$e^x = \text{the limiting value of } \left(1+\frac{x}{n}\right)^n \text{ as } n \text{ approaches infinity}$$

Q3 Use the definition of e^x and a large n to approximate the value of e^3.

SKETCH AND INVESTIGATE

Now that you know how to approximate e^x when x is a real number, you're ready to consider $e^{i\pi}$. Raising e to an imaginary power certainly seems strange, but let's use our definition of e^x with $x = i\pi$ and see what happens:

$$e^{i\pi} = \text{the limiting value of } \left(1+\frac{i\pi}{n}\right)^n \text{ as } n \text{ approaches infinity}$$

As before, we can get a sense of how this expression behaves by starting with a small value of n. When $n = 10$, we must evaluate $\left(1+\frac{i\pi}{10}\right)^{10}$.

4. Open the sketch **eipi.gsp** in the **7 Polar Coords and Complex Nos** folder. The axes represent the complex plane with real numbers on the horizontal axis and imaginary numbers on the vertical axis. Point A is at $(1, 0)$ and represents the value 1. Point B is at $\left(1, \frac{\pi}{10}\right)$ and represents the value $1 + \frac{i\pi}{10}$.

This is a good time to review the activity Multiplication of Complex Numbers.

Q4 The sketch provides two pieces of information about right triangle OAB: the measure of $\angle AOB$ and the length of OB. Describe geometrically what it means to multiply the complex number $1 + \frac{i\pi}{10}$ by itself. Describe geometrically what it means to raise $1 + \frac{i\pi}{10}$ to the tenth power.

Sketchpad makes the process you described in Q4 simple to carry out.

5. Select point A and $n - 1$, and hold down the Shift key. Choose **Transform | Iterate to Depth.**

6. Click on point B to map point A to point B. Click *Iterate* to confirm your mappings. The iterated triangles appear, all of which are similar to $\triangle OAB$.

Q5 Identify the locations of $\left(1 + \frac{i\pi}{10}\right)^2, \left(1 + \frac{i\pi}{10}\right)^3, \ldots, \left(1 + \frac{i\pi}{10}\right)^9$, and $\left(1 + \frac{i\pi}{10}\right)^{10}$.

7. Select the iterated point image and choose **Transform | Terminal Point.** Label the terminal point P.

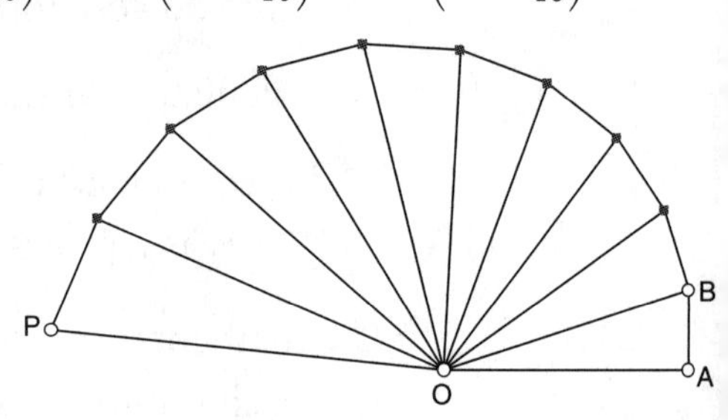

8. With point P selected, choose **Measure | Coordinates.**

Q6 What is the value of $\left(1 + \frac{i\pi}{10}\right)^{10}$?

The third page of the sketch provides an intuitive explanation of what you're observing.

9. Drag slider point n to the right to increase the value of n.

Q7 What do you notice about the value of $\left(1 + \frac{i\pi}{n}\right)^n$ as n grows larger?

EXPLORE MORE

You can generalize the method of finding $e^{i\pi}$ to compute e raised to the $i\theta$ power, where θ is any number.

10. Open page 2 of **eipi.gsp.**

This sketch computes e raised to the power $\frac{i\pi}{k}$ for any value of k.

11. Double-click the parameter k and change its value to 2.

Q8 Use the iteration process from page 1 of the sketch to approximate the value of $e^{i\pi/2}$.

Q9 Approximate the imaginary powers of e for values of k such as 3 and 4.

Q10 For each of your approximations, what do you notice about its distance from the origin?

Q11 When $k = 2$, what angle does the point representing $e^{i\pi/2}$ make with the x-axis? Answer this question for $k = 3$ and for $k = 4$.

Q12 The eighteenth-century mathematician Leonard Euler developed the identity $e^{i\theta} = \cos\theta + i\sin\theta$ as a way to compute the value of e raised to any imaginary power. Explain how this identity makes sense based on your answers to Q10 and Q11.

Q13 Substitute $\theta = \pi$ into the identity in Q12. What do you get?

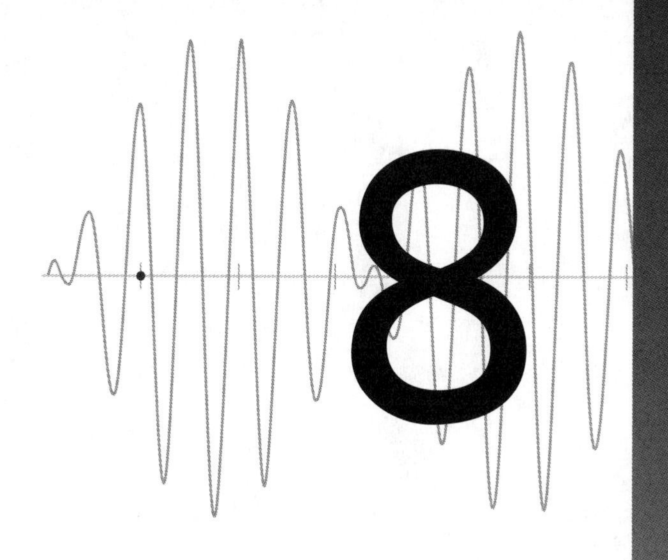

8

Sequences and Series

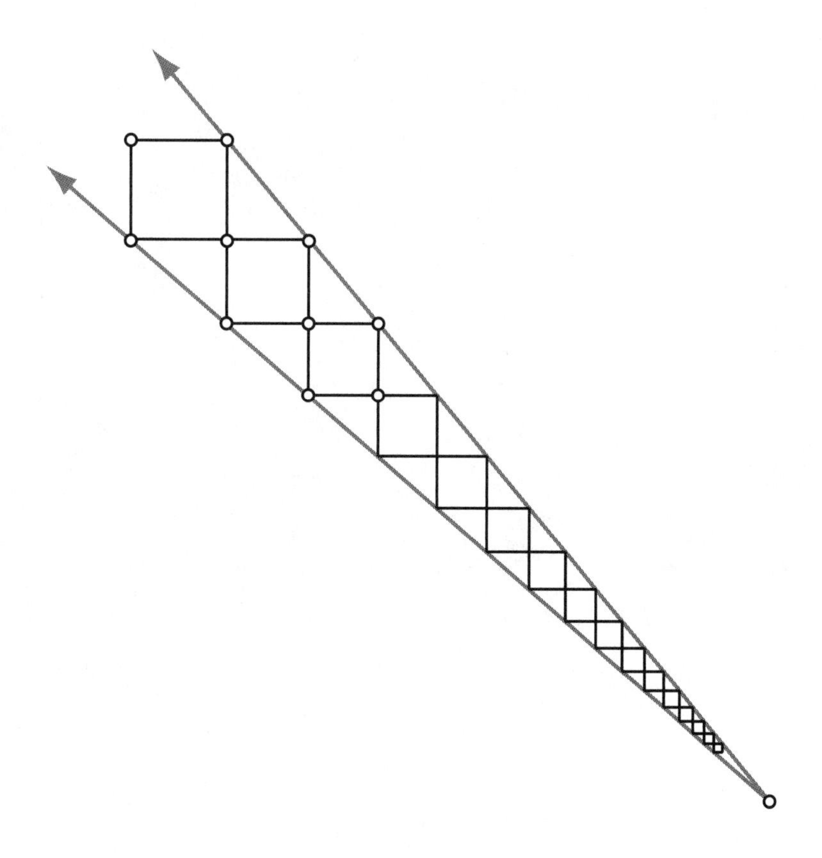

Generating Arithmetic and Geometric Sequences Numerically

In this activity, you'll build and explore arithmetic and geometric sequences by using Sketchpad's iteration feature.

ARITHMETIC SEQUENCES

Open **Sequences.gsp** in the **8 Sequences and Series** folder. This sketch includes a start value of 2 and a *difference* of 3. With these two values, you can generate an arithmetic sequence.

Q1 Look at the number line on the sketch. What arithmetic sequence is shown? How do the numbers in the sequence relate to the *start* and *difference* values?

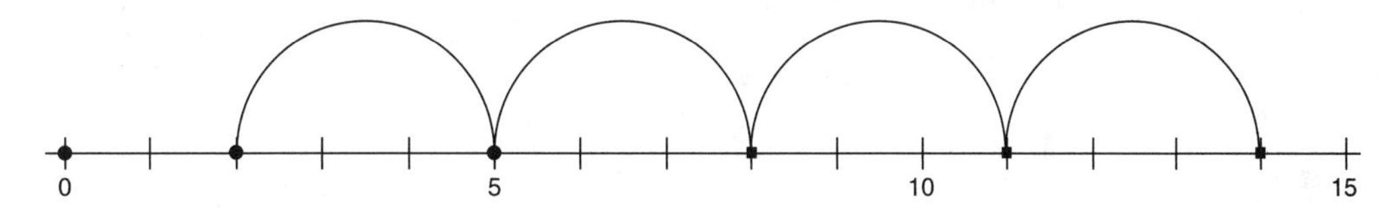

Now you'll create a table of values that corresponds to the arithmetic sequence on the number line.

1. Choose **Measure | Calculate** to display the Calculator. Click on *start* in the sketch, the + sign on the keypad, and *difference* in the sketch to compute *start* + *difference.*

ne beginning value for n iteration is often alled the *seed,* or ne *pre-image,* of ne iteration.

2. Select *start,* and choose **Transform | Iterate.** Map *start* to *start* + *difference* by clicking on *start* + *difference.* Then click Iterate to confirm the mapping.
3. A table appears with the 2nd through 5th terms in your arithmetic sequence. To increase the number of terms in your sequence, select the table and press the + key on your keyboard several times. You can decrease the number of terms by pressing the − key.

ɔ change the value of a arameter, double-click with the **Arrow** tool nd enter a new number.

Q2 Your sequence does not include the term 24. Find two ways to change the sequence so that it includes 24.

Q3 Below are several arithmetic sequences. For each one, find the *start* and *difference* values that generate them.

a. 3, 6, 9, 12, 15, . . .

b. −10, −14, −18, −22, −26, . . .

c. 1, 1, 1, 1, 1, . . .

d. 0.5, 0.75. 1.0, 1.25, 1.5, . . .

nswer this question ithout creating the equence.

Q4 Suppose the *start* value of your sequence is 4 and the *difference* is 6. Will there be a term in your sequence between 2000 and 2010?

GEOMETRIC SEQUENCES

Page 2 of **Sequences.gsp** includes a *start* value of 1 and a *ratio* of 3. With these two values, you can generate a geometric sequence.

4. Using the directions for creating an arithmetic sequence as a guide, create a table that corresponds to the geometric sequence shown on the number line.

Q5 Your sequence does not include the term 24. Describe two ways to change the sequence so that it includes 24.

Q6 Below are several geometric sequences. For each one, find the *start* and *ratio* values that generate them.

a. 2, 8, 32, 128, 512, . . .

b. 32, −16, 8, −4, 2, . . .

c. 1, 1, 1, 1, 1, . . .

d. 1, −1, 1, −1, 1, . . .

Q7 Change your sequence so that *start* = 1 and *ratio* = 3. How many copies of the 2nd arc (between 3 and 9) can fit into the 3rd arc? How many copies of the 3rd arc can fit into the 4th arc? Does this pattern continue?

EXPLORE MORE

Hint: Your iteration requires using two pre-image parameters.

Q8 The Fibonacci sequence begins 0, 1, 1, 2, 3, 5, 8, 13, 21, . . . , where each term is the sum of the preceding two terms. Use the seed values on page 3 of the sketch to generate an iterated table of Fibonacci values.

Area Models of Geometric Series

In this activity, you'll use various methods to dissect a square and create area models for certain geometric series. The models reveal in beautiful ways the values to which the series converge.

DISSECTION 1

Open **Area Models.gsp** in the **8 Sequences and Series** folder. You'll see a quadrilateral, *ABCD,* and two parameters, *depth* and *sum.* You will use quadrilateral *ABCD* to find the sum $\frac{1}{2} + \frac{1}{4} + \frac{1}{8} + \ldots.$

1. Construct midpoint *E* on segment *AB* and midpoint *F* on segment *CD.*

2. Construct quadrilateral *EBCF.*

Q1 What fraction of the entire figure does this new quadrilateral represent?

) choose the **Relative** **rea** tool, press and hold n the **Custom** tool icon t the bottom of the oolbox. Choose the ol from the menu at appears.

3. Click the **Relative Area** custom tool on the quadrilateral to measure its area relative to the area of *ABCD.* Switch back to the **Arrow** tool. If the measured area is not exactly 0.50000, click the *Calibrate Areas* button.

4. Use Sketchpad's Calculator to calculate *sum* + *area.*

5. Select points *A, B,* and *C,* parameter *sum,* and parameter *depth.* While pressing the Shift key, choose **Transform | Iterate To Depth.**

) map *A* ⇒ *E,* click on oint *E* in the sketch. point *E* is behind the alog box, move the alog box aside.

6. In the Iterate dialog box, map *A* ⇒ *E, B* ⇒ *F, C* ⇒ *D,* and *sum* ⇒ *sum* + *area.* The dialog box should appear as shown here, and the figure and table should appear as shown at right. Click Iterate to confirm your mappings.

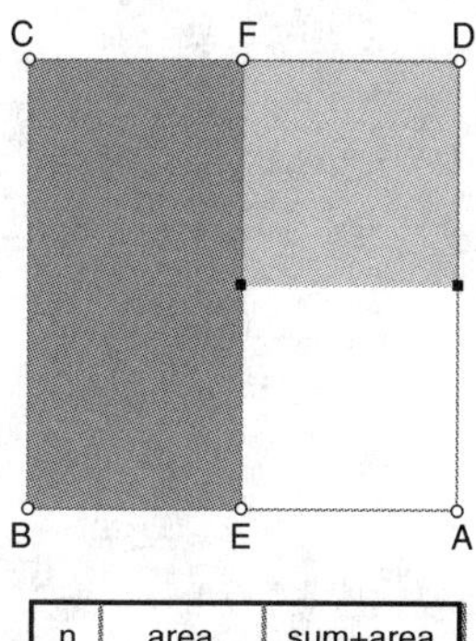

n	area	sum+area
0	0.50000	0.50000
1	0.25000	0.75000

Iterate

Pre-Image	To	First Image
A	⇒	E
B	⇒	F
C	⇒	D
sum	⇒	sum+area

Number of iterations: 1.

Display | Structure | ? | Cancel | Iterate

7. Select the *depth* parameter, and press the **+** key on the keyboard several times to increase the depth of iteration.

Q2 How much of the original figure will the iterated quadrilaterals eventually fill?

Q3 What depth of iteration is required for the partial sum (the accumulated area of the iterated polygons) to become greater than 0.99999?

Q4 What depth of iteration do you think is required for the partial sum to be exactly 1?

DISSECTION 2

On page 2 of the sketch, you'll investigate the sum $\frac{1}{3} + \frac{1}{9} + \frac{1}{27} + \ldots$.

8. Construct the polygon *BEFGHI.*

Q5 What fraction of the total area does this polygon represent?

9. Press the *Hide Segments* button. Then select *A, B, C,* and *depth.* While holding the Shift key, choose **Transform | Iterate To Depth.**

10. In the Iterate dialog box, map $A \Rightarrow A$, $B \Rightarrow E$, and $C \Rightarrow F$. Press the – key twice so that only a single iteration appears in the sketch.

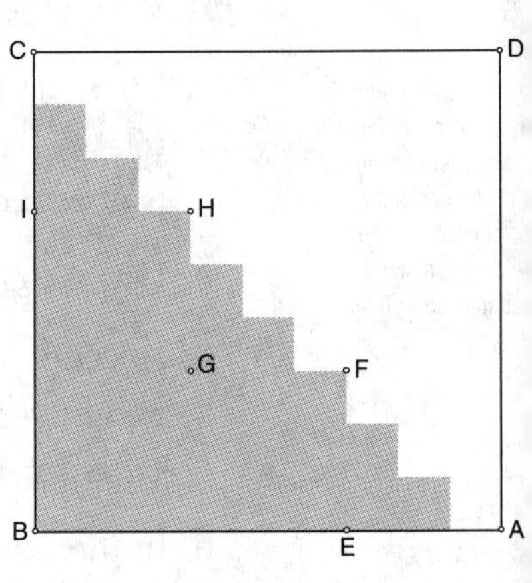

11. Choose **Add New Map** from the Structure pop-up in the dialog box, and map $A \Rightarrow F$, $B \Rightarrow G$, and $C \Rightarrow H$. Choose **Add New Map** again, and map $A \Rightarrow H$, $B \Rightarrow I$, and $C \Rightarrow C$. Click Iterate to confirm your mappings.

12. Hide the segments that appear as part of the iteration so that only the iterated polygons remain.

Q6 What fraction of the total area do the three new polygons represent? Justify your answer.

Q7 Increase the depth from 1 to 3 or 4. What fraction of the entire figure do all the polygons take up?

DISSECTION 3

On page 3, you'll use a different construction to investigate the sum $\frac{1}{3} + \frac{1}{9} + \frac{1}{27} + \ldots$

13. Construct polygons *AEHD* and *FBCG,* and make them contrasting colors.

Q8 What fraction of the total area does each polygon represent?

14. Click the *Calibrate Areas* button, and then click the **Relative Area** custom tool on polygon *AEHD* to measure its area relative to the area of *ABCD.* Switch back to the **Arrow** tool.

15. Calculate *sum* + *area.*

16. Select *A, B, C, sum,* and *depth,* and choose **Transform | Iterate To Depth.** Map $A \Rightarrow F$, $B \Rightarrow G$, $C \Rightarrow H$, and *sum* ⇒ *sum* + *area.* After you confirm the mappings, hide the iterated images of the segments and points, so only the iterated polygons remain.

Q9 When you increase the depth, what fraction of the total figure does each iterated polygon occupy? What does this mean about the partial sum of the geometric series?

Q10 To what depth must you iterate so that the value of *sum* + *area* is greater than 0.49999?

EXPLORE MORE

After you change a shape, be sure to click the *Calibrate Areas* button if one exists on the page.

Q11 Must the shapes used on pages 1, 2, and 3 of the sketch be squares? Try changing the shape of quadrilateral *ABCD* on each of these pages. What effect (if any) does this have on your results?

17. Go to page 4 of the sketch, which shows a dissection of a square into five pieces.

Q12 What fraction of the area of the square does each piece represent? Justify your answer.

18. Construct polygons covering each of the four outer pieces, coloring them differently.

19. Measure the relative area of one of the polygons. Then iterate the construction into the inner piece.

Q13 What geometric series is represented by this figure? What is the sum of this series? Explain your answers.

20. Go to page 5 of the sketch, which shows a dissection of a square into four pieces.

Q14 What fraction of the area of the square does each piece represent? Justify your answer.

Construct polygons covering each piece that is not a square, coloring them differently. Measure the relative area of one of the polygons. Then iterate the construction into the square piece.

Q15 What geometric series is represented by this figure? What is the sum of this series? Explain your answers.

Q16 What conclusion can you draw concerning the sum of the series $\frac{1}{n} + \frac{1}{n^2} + \frac{1}{n^3} + \ldots$? Explain your answer geometrically.

A Geometric Series Coil

In a geometric sequence, each term is obtained from the previous term by multiplying by a constant ratio, r. A geometric series is simply the sum of these terms. If the first term of a series is a and its ratio is r, the series is

$$a + ar + ar^2 + ar^3 + \ldots$$

In this activity, you'll build a coil that contains multiple segments. The lengths of these segments will form terms in a geometric series.

CREATE THE COIL

Open the sketch **Geometric Coil.gsp** in the **8 Sequences and Series** folder. You'll begin this activity by constructing a rotated dilated image of point A. You can do this by following either the general instructions or the detailed instructions.

General Instructions

Rotate segment AB about point B by $\angle XYZ$, and then dilate it by the measurement labeled *ratio.* Label the resulting point C. Construct segment BC. Drag point Z, and observe what happens to $\angle ABC$.

Detailed Instructions

1. Select, in order, points X, Y, and Z that form $\angle XYZ$. Choose **Transform | Mark Angle.**
2. Select point B, and choose **Transform | Mark Center.**
3. Select point A, and choose **Transform | Rotate.** Rotate point A by the marked angle XYZ. Label the rotated point A'.
4. Select the value labeled *ratio,* and choose **Transform | Mark Scale Factor.**

The ratio is equal to r_1/r_2.

5. Select point A', and choose **Transform | Dilate.** Dilate point A' by the marked ratio. Label the dilated point C.
6. Construct segment BC. Hide A'. Drag point Z, and observe what happens to $\angle ABC$.

ITERATE THE COIL

Now that you've constructed segment BC through a combination of a rotation and a dilation, you'll apply this same geometric recipe over and over to create successive pieces of the coil. Sketchpad's iteration feature is perfect for the job.

Each segment of the coil will be obtained from the previous segment through a rotation by $\angle XYZ$ and a dilation by 0.75.

7. Select points A and B, and choose **Transform | Iterate.** In the Iteration dialog box, designate point B as the image of point A by clicking on point B in the sketch. Similarly, designate point C as the image of point B.

8. Click Iterate in the dialog box to iterate the segment three times.

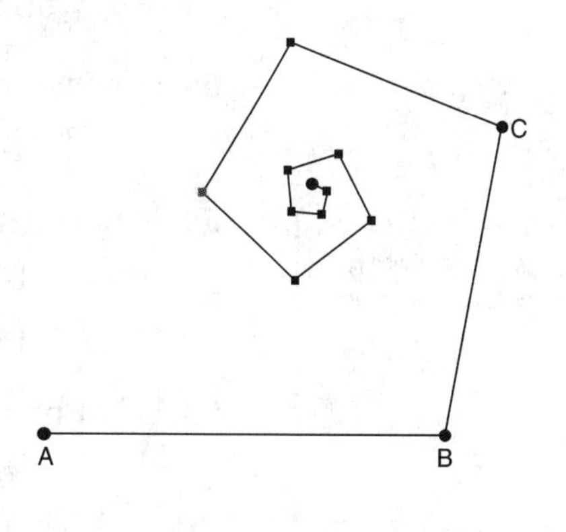

9. With the iterated image selected, press the + key on your keyboard several times to increase the number of iterations.

10. Click on any vertex of the iterated image to select all points of the iteration. Choose **Transform | Terminal Point** to create the endpoint of the iterated coil. Label this point S.

11. Slowly drag point Z, and watch the coil wind and unwind.

12. Press the *Make angle 180 degrees* button to unwind the coil completely and lay it flat.

SUM THE SERIES

Write your answers in a form that includes 0.75.

Q1 In your coil, segment $AB = 1$ and segment $BC = 1 \cdot 0.75$. What are the lengths of the next three segments? Assuming that you could add an infinite number of segments to the coil, what geometric series would your unfolded coil represent?

13. Select point S, and choose **Measure | Abscissa (x).**

Q2 What is the current sum of your geometric series?

14. Select your iterated image, and add more segments to it by repeatedly pressing the + key on your keyboard.

Q3 Imagine that you could add an infinite number of segments to your coil. As the total length became larger and larger, what value would it approach?

CREATE A TABLE OF VALUES

Now that you've built your coil, you'll create a table of values that shows the lengths of its individual segments and the sums of these lengths as the coil grows in size.

15. Use Sketchpad's Calculator to calculate *sum* + *a*. This is the sum of the geometric series when there is only one term—namely, *a*.

16. Calculate the 2nd term of the geometric sequence, $a \cdot ratio$.

17. Select both the a and sum parameters and, using iteration, map $a \Rightarrow a \cdot ratio$ and map $sum \Rightarrow sum + a$.

18. Select the resulting table, and increase the depth of iteration by pressing the + key on the keyboard.

Q4 How many terms does it take before the sum of the geometric series is greater than 3.99?

Double-click the parameters r_1 and r_2 to change their values.

Q5 Experiment with different infinite series by varying the value of the ratio. Try $ratio = 1/3$ and $ratio = 4/5$, among other values. Based on the value of the ratio, can you predict what the sum will be?

Q6 What happens when r is greater than or equal to 1?

A Geometric Series Staircase

In this activity, you'll explore a visual representation of an infinite geometric series. By investigating the geometry of this model, you'll develop a formula for summing any infinite geometric series.

CREATE A STAIRCASE

Open **Geometric Staircase.gsp** in the **8 Sequences and Series** folder. You'll see a square, *BCDE*, with side length *a*, and a parameter labeled *r* (for "ratio"). Follow the steps to create a staircase of squares.

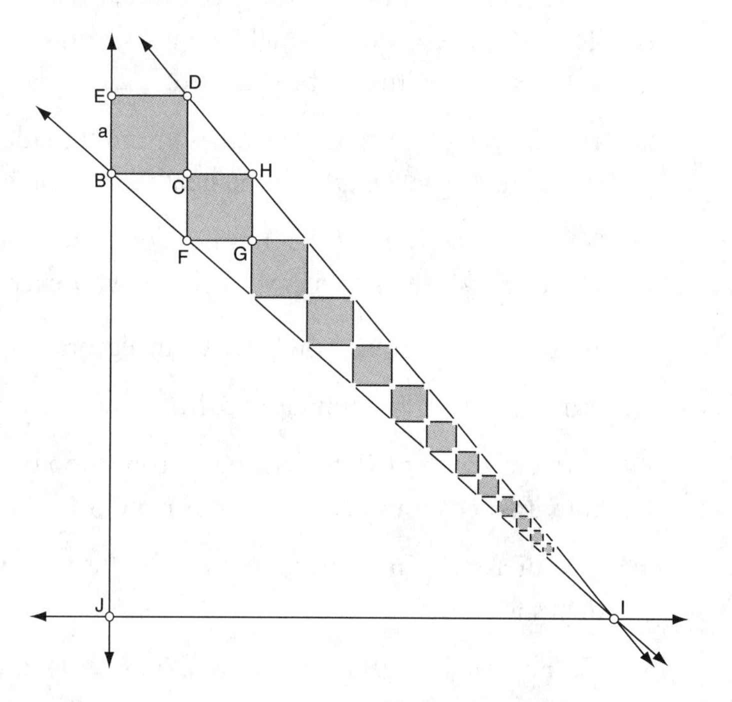

Select the square, and choose **Transform | Dilate.** With the Dilate dialog box open, click on parameter *r* in the sketch.

1. Dilate the square (including its vertices and edges) about point *C* by the parameter *r*.
2. Rotate the dilated square by 180°.
3. Hide the original dilated square.
4. Label the three new vertices as shown in the diagram.
5. Iterate point *B* to point *F* and point *C* to point *G*.

Select *B* and *C* and choose **Transform | Iterate.** With the Iterate dialog box open, click on points *F* and *G* in the sketch.

The iterated squares and a table with iterated values of *a* are created simultaneously.

6. With the iterated image of the square selected, press the + key on the keyboard repeatedly to add more squares to your staircase.
7. Construct a line through points *B* and *F* and another line through points *D* and *H*. Construct the intersection point of these lines, and label it point *I*.

8. Construct a line through points *E* and *B*.

9. Construct a line through point *I* perpendicular to the line through points *E* and *B*. Construct the intersection point of the two lines, and label it point *J*.

SUM THE SERIES

Notice that your iterated squares all sit between two lines: the line through points *B* and *F* and the line through points *D* and *H*. Assuming that you could iterate your original square an infinite number of times, the squares would grow smaller and smaller and converge to the single point *I*. Contained in this shrinking staircase is a geometric series waiting to be summed.

Q1 The side length of square *BCDE* is *a*, and the side length of square *FGHC* is *ar*. What are the sidelengths of the next four squares in the staircase?

Q2 Let *S* equal the sum of the infinite geometric series $a + ar + ar^2 + ar^3 + \ldots$. There are two segments in your sketch whose length is *S*. Which two are they?

In the next few questions, you'll derive an algebraic formula for finding the value of *S*.

Hint: What is the sum of $\angle CBF$ and $\angle IBJ$?

Q3 Prove that $\triangle BCF$ is similar to $\triangle IJB$.

Q4 Find the length of $\overline{JB}$ by writing a proportion based on the similar triangles. Your answer should be in terms of *r* and *S*.

Q5 In your sketch, $JB + BE = JE$. Rewrite this equality in terms of *S*, *r*, and *a*, and then solve for *S*.

Remember, *S* is the sum of the series.

10. Use Sketchpad to calculate *S* based on your formula from Q5. When forming your equation, click directly on *a* and *r* so that the value of *S* will update itself automatically when you change these values.

To check your formula, measure directly a segment in your sketch whose length is *S*.

You can find the sum of a different series by dragging point *B* to change the *a* value or by double-clicking *r* to change the value of the ratio.

Q6 Choose a value of *r* greater than or equal to 1. What happens to your staircase? Does the series still converge to a finite sum?

aylor Series

You can compute the value of a polynomial function directly and easily for any particular value of x using multiplication and addition. But values of other functions, such as the sine function, are much more difficult to compute.

u can learn how to rive the Taylor series nen you study calculus.

In this activity, you'll approximate the sine function using a series called a *Taylor series* and observe the behavior of the partial sums when the series is evaluated to various depths. The Taylor series approximation for sin(x) is

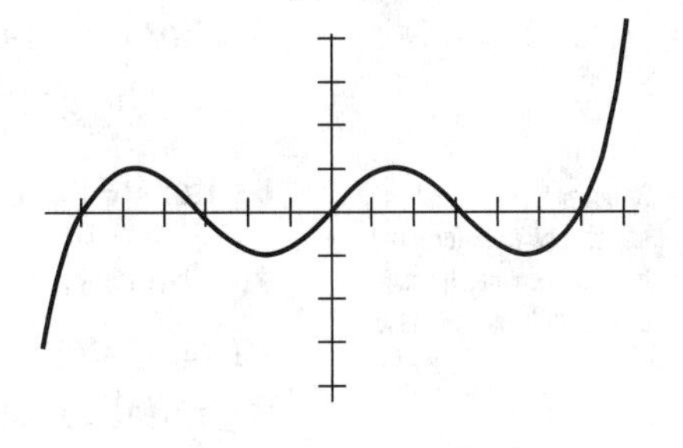

$$f(x) = \frac{x^1}{1!} - \frac{x^3}{3!} + \frac{x^5}{5!} - \frac{x^7}{7!} + \frac{x^9}{9!} - \frac{x^{11}}{11!} + \ldots$$

SKETCH AND INVESTIGATE

1. In a new sketch, create a square grid, construct point A on the x-axis, and measure the point's x-value. Label the x-value x.

2. Create four parameters to use in iterating the series. Label them i, *num*, *den*, and *sum*.

Q1 Parameter i represents the index for the terms, following the sequence 1, 3, 5, ... What rule can you apply to one element of this sequence to calculate the next?

Q2 Parameter *num* represents the numerator, taking on values x, $-x^3$, x^5, $-x^7$, and so forth. What is the rule to calculate a value of this sequence from the previous value?

Q3 Parameter *den* represents the denominator, taking on values 1!, 3!, 5!, and so forth. What's the rule to calculate the next value of this sequence? (Express your answer in terms of the previous values of *den* and i.)

Q4 Parameter *sum* represents the sum of all the terms from the first term through the ith term. What value should you use as the initial value of the sum, before adding the very first term? What's the rule to calculate one sum from the previous sum?

arameters must be dependent values in der to be iterated, so ou can't set the initial alues of the parameter at depends on x until ter you construct the eration.

3. All but one of these parameters have constant initial values that you can assign now. (The initial value for the other isn't constant, but depends on the value of x.) Assign appropriate initial values to the parameters that don't depend on x. Assign an initial value to the other parameter as though the value of x were 2.

Q5 What initial values did you assign to the parameters?

Make sure the calculated values are what you expect.

4. For each of the four parameters, use the rule you described above to calculate th next value of the quantity it represents. (Your calculations should involve only the values of the four parameters and the value of x.)

5. Plot the point (x, *sum*). The iterated image of this plotted point will allow you tc see the graph of each successive expansion of the Taylor series.

Select all four parameters, and choose **Transform | Iterate.** Then match each parameter to its next value.

6. Iterate each of the parameters to its calculated next value.

7. The parameter *num* doesn't yet have a correct initial value, because the initial value depends on x. Select *num*, choose **Edit | Edit Parameter,** and calculate the initial value so that it depends correctly on x.

8. Drag point A left and right on the x-axis, observing the values in the table and the positions of the plotted points.

The last row of the table should contain $n = 1$.

9. Select the iterated image of the plotted point and press the – key on the keyboard twice to set the depth of iteration to 1.

The *terminal point* is the very last image of the iterated point, based on the current depth of iteration.

10. With the iterated image of the plotted point still selected, choose **Transform | Terminal Point.** Then construct the locus of the terminal point as A moves along the axis.

Q6 What is the shape of the locus? Which terms contribute to this shape?

To increase the depth, select either iterated image (the table or the image of the plotted point), and press the + key on the keyboard.

11. Set the depth of iteration to 2.

Q7 How does this change the shape of the locus? Which terms contribute now?

12. Increase the depth to 3. Turn on tracing for both the iterated image of the plotted point and the locus. Animate point A, and observe the behavior of the point images.

n	i+2	-num·x²	den·(i+1)·(i+2)	sum+$\frac{num}{den}$
0	3.00	-8.00	6.00	2.00
1	5.00	32.00	120.00	0.67
2	7.00	-128.00	5040.00	0.93
3	9.00	512.00	362880.00	0.91

Q8 What shapes do the iterated point images trace? Sketch their shapes and explain the role of each trace based on the terms of the series.

You may have to move the origin and change the domain of the locus to see two full periods.

13. While point A is moving, increase the depth until the locus accurately approximates the sine curve for at least two periods.

Q9 How many terms are required to give a reasonable approximation for the first period of the sine function? For the first two periods?

EXPLORE MORE

The Taylor series for the cosine function is as follows:

$$f(x) = \frac{x^0}{1} - \frac{x^2}{2!} + \frac{x^4}{4!} - \frac{x^6}{6!} + \frac{x^8}{8!} - \frac{x^{10}}{10!} + \cdots$$

You can change the initial values of your existing iteration to calculate this series. Decide which parameters to change, and calculate and plot the modified series.

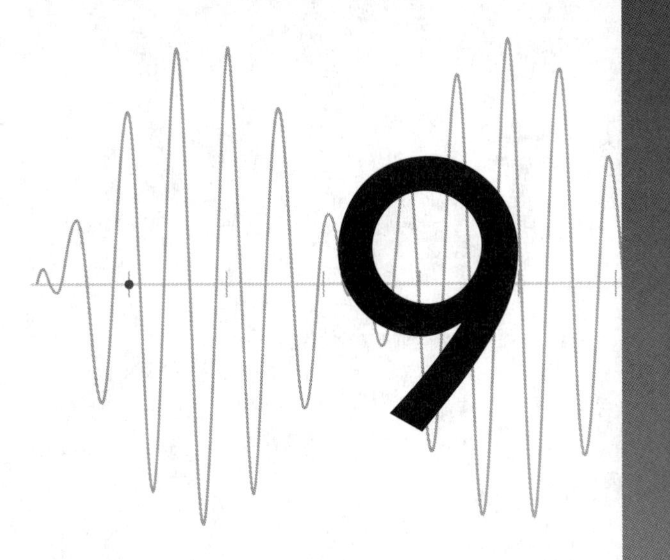

Introduction to Calculus

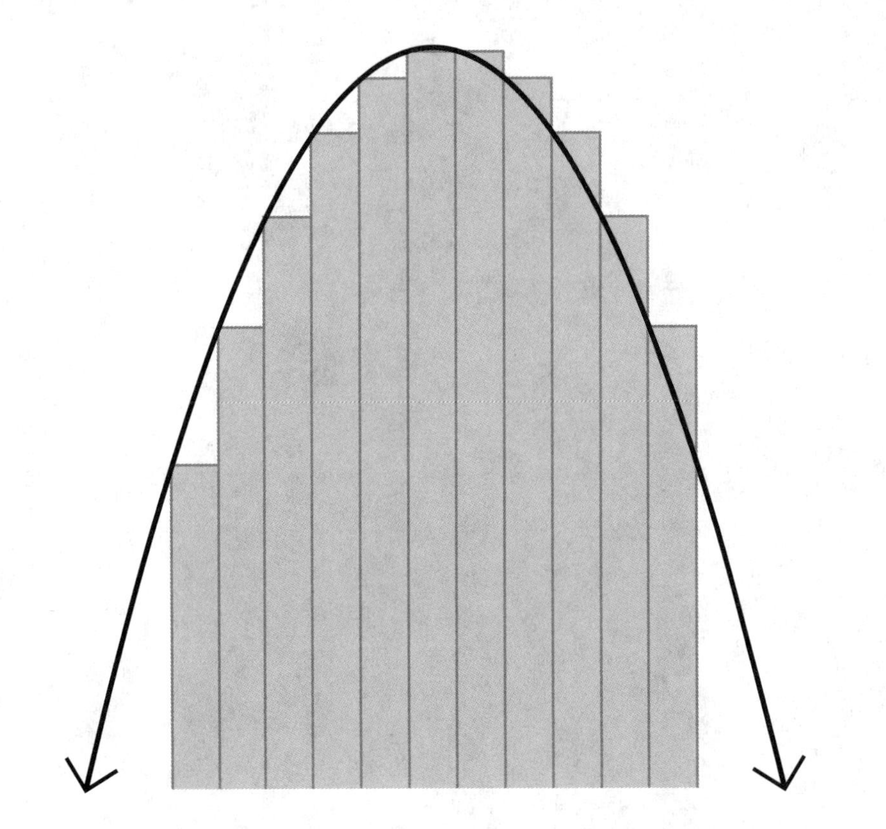

Instantaneous Rate

Many schools have doors equipped with automatic closers. When you push such a door, it opens quickly, and then the closer closes it again, more and more slowly until it finally closes completely.

DOOR ANGLE AS A FUNCTION OF TIME

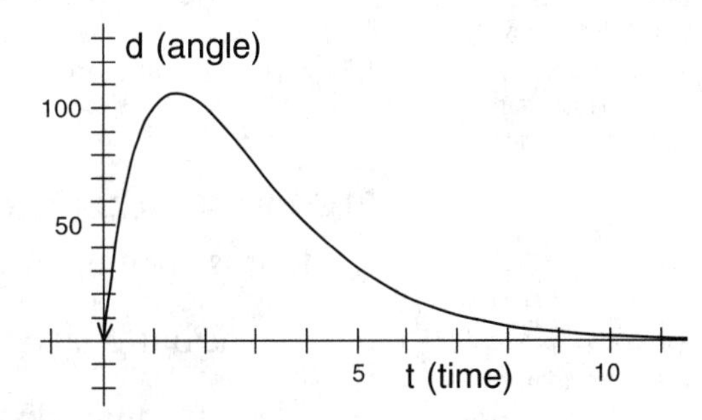

1. Open the sketch **Instantaneous Rate.gsp** from the **9 Introduction to Calculus** folder. Press the *Open Door* button to operate the door. The door opens and closes, and the graph shows the angle of the door (d) in degrees as a function of time (t) in seconds.

2. Drag point t_1 back and forth along the time axis, and watch how the angle of the door changes and how the point on the graph corresponds to the door's angle. Observe the values of the t_1 and d_1 measurements as you drag.

Q1 For what values of t_1 is the angle increasing? How can you tell?

Q2 What is the maximum angle the door reaches? At what time does this occur?

THE DOOR AT TWO DIFFERENT TIMES

The value Δt is the separation between the two values of time (t_1 and t_2).

3. To find the rate of change of the door's angle, you need to look at the door's position at two different times. Press the *Show t2* button to see a second point on the time axis slightly separated from point t_1. Drag point t_1 back and forth, and observe the behavior of the new points on the graph. To change the separation of the two times, press the button labeled *1.0* and then the button labeled *0.1*.

4. Make the separation of the two points smaller than 0.1. Can you still see two distinct points on the graph? Can you see the values of t_2 and d_2 change as you make Δt smaller? Experiment with dragging the Δt slider, to change the separation of the two values of time directly.

Q3 What is the largest separation you can get by moving the slider? What is the smallest separation you can actually observe on the graph?

Q4 As you make Δt smaller, can you observe changes in the numeric values of t_2 and d_2 even when you can no longer observe any changes on the graph?

THE RATE OF CHANGE OF THE DOOR'S ANGLE

Hint: Divide the change in the angle by the change in the time.

5. Set Δt to 0.1, and then use the numeric values of t_1, d_1, t_2, and d_2 to calculate the rate of change of the door's angle at any particular time. (Use Sketchpad's Calculator to perform this calculation.)

Q5 What are the units of the rate of change? What does the rate of change tell you about the door's motion?

When you press the *Show Rate* button, a dotted line appears connecting the two points on the graph.

6. Press the *Show Rate* button to check your result.

Q6 What is the relationship between the dotted line and the rate of change you calculated?

Q7 Move t_1 back and forth. How can you tell from the rate of change whether the door is opening or closing? How can you tell whether its rate is fast or slow?

After double-clicking the table, it shows two rows of numbers, with the first row permanent and the second row changing as the measurements change.

Q8 Use the buttons to set t_1 to 1.0 and Δt to 0.1. What is the rate of change?

7. Select the numeric values of t_1, d_1, t_2, d_2, Δt and the rate of change. With these six measurements selected, choose **Graph | Tabulate.** Double-click the table to make the current entries permanent.

THE LIMIT OF THE RATE OF CHANGE

You may want to press the *0.1* button and then the *0.01* button again to check the motion of the dotted line.

8. Set the time interval (Δt) to exactly 0.01. Note the new value of the rate of change. Could you see the dotted line move as you reduced the time interval? Double-click the table to permanently record these new values.

Q9 How does this rate of change compare to the value when Δt was 0.1?

9. Similarly, record in the table values for time intervals of 0.001, 0.0001, 0.00001, and 0.000001.

Q10 What happens to the value of the rate of change as the time interval becomes smaller and smaller? What value does the rate of change appear to approach?

Q11 Can you see the dotted line move as Δt changes from 0.001 to 0.0001?

10. Set the value of t_1 to 3 seconds (by pressing the *t->3* button), and collect more data on the rate of change of the door's angle. Collect one row of data for each time interval from 0.1 second to 0.000001 second.

The *average rate of change* is the rate of change between *two* different values of *t*. The *instantaneous rate of change* is the exact rate of change at *one* specific value of *t*. Because you must have two different values to calculate the rate of change, one way to measure the instantaneous rate of change is to make the second value closer and closer to the first and then find the *limit* of the average rate of change as the interval gets very small.

The instantaneous rate of change of a function—that is, the limit of the average rate of change as the interval gets close to zero—is called the *derivative* of the function.

Q12 What is the derivative of the door's angle when t_1 is 3 seconds?

One Type of Integral

Consider a car starting up from a red light. Its velocity increases for a while and then levels off at 60 ft/s. From a graph of the velocity, how could you determine the distance the car has traveled in a given period of time?

e distance traveled om $t = 70$ s to $= 100$ s is equal to e area of the shaded ctangle on the right, cause *rate* × *time* is ual to *height* × *width*.

Recall that *distance* = *rate* × *time*. The velocity is constant in the interval on the far right, from $t = 70$ s to $t = 100$ s, so you can multiply rate by time (60 ft/s × 30 s) to find that the car travels 1800 ft during this period of time.

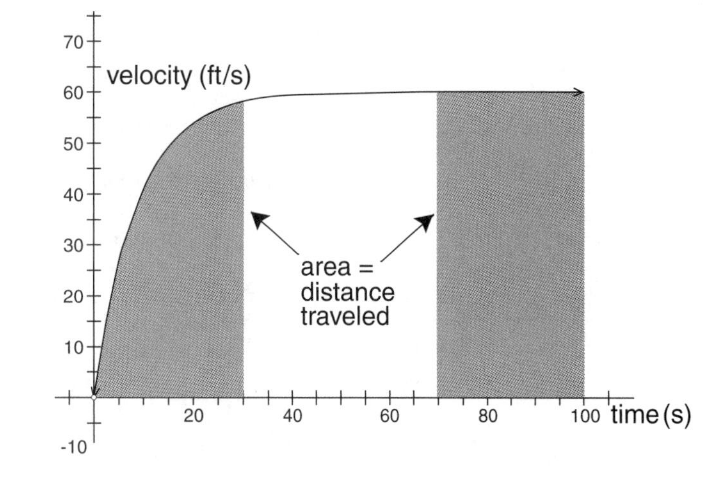

e definite integral of a nction corresponds to e area under the graph the function.

Finding how far the car travels during the time from $t = 0$ s to $t = 30$ s is harder, because the velocity is changing. The process of finding this distance, when the velocity is changing, is called finding the *definite integral* of the velocity.

THE DISTANCE A CAR TRAVELS

1. Open the sketch **Definite Integral.gsp** from the **9 Introduction to Calculus** folder. Press the *Drive Car* button to start the car. Observe how the car's velocity behaves, starting from zero and ending at 60 ft/s.

2. To find the total distance the car travels during any period of time, you'll have to estimate the area under the curve. Press the *Show Grid* button to display a grid you can use to estimate the area.

Q1 How wide is each square of the grid? What are the units?

Q2 How high is each square of the grid? What are the units?

Q3 What is the area of each square of the grid? What are the units?

Q4 How many squares are in the shaded region on the right? How can you use this result to confirm the distance traveled from $t = 70$ s to $t = 100$ s?

Now you'll count squares to find the area for the shaded region on the left.

keep track, you can e the **Point** tool to put oint inside each uare you count.

3. Count the number of complete squares totally contained within this region.

4. Now estimate the area of the squares that are partially within the shaded region. For each such square, estimate whether it is more than half shaded or less than

half shaded. Count only the squares that are more than half shaded. Add this number to the number of complete squares you counted in step 3.

Q5 What is your estimate of the number of squares in the shaded region on the left?

Remember to multiply the number of squares by the value represented by each square.

Q6 What is the distance the car traveled from $t = 0$ s to $t = 30$ s?

5. Select the *Square Size* parameter, and then press the – key on the keyboard to change the size of the squares to 2.0.

Count the complete squares first, and then count all the partial squares that are more than half shaded.

6. Count the squares in the shaded region on the left.

Q7 How many squares do you get this time? What is the area of each square?

Q8 Based on this count, what is the distance the car traveled from $t = 0$ s to $t = 30$ s?

Q9 Do you think this estimate is more or less accurate than the previous one? Why? How could you make it still more accurate?

Q10 Estimate the area from $t = 30$ s to $t = 70$ s. Add your results for the three areas to find the total distance the car traveled in 100 s.

DEFINITE INTEGRALS FOR OTHER FUNCTIONS

7. On page 2 of the sketch, you'll find another function, $y = 8 \cdot 0.7^x$. Estimate the definite integral for this function, using the domain from $x = 1.00$ to $x = 7.00$.

Q11 What is your estimate of the definite integral for this function?

Q12 Double-click the **Arrow** tool on the *Square Size* parameter, and change the value of the parameter to 0.5. What is your new estimate of the definite integral for this function?

Q13 Change the value of the parameter to 0.1. If you had to count such small squares what kind of function would you prefer to have? Why?

EXPLORE MORE

Be careful to determine the area of each rectangle correctly.

On page 3 of the sketch, you can set both the height and width of the rectangles. Use several different-size rectangles to estimate the definite integral for this function.

Using the grid on page 4 of this sketch, plot some other functions of your choice and estimate the definite integrals on the specified intervals. Here are some possible choices.

What should you do about cells below the *x*-axis?

$$f(x) = \sin(x) \text{ from } x = 0 \text{ to } x = \pi$$

$$f(x) = \sin(x) \text{ from } x = \pi \text{ to } x = 3\pi/2$$

$$g(x) = x^2 - 2x - 1 \text{ from } x = 1 \text{ to } x = 2$$

$$g(x) = x^2 - 2x - 1 \text{ from } x = 2 \text{ to } x = 3$$

Rectangular and Trapezoidal Accumulation

As you've discovered, it's time consuming to use squares to accurately estimate a definite integral. An accurate estimate requires small squares, so there are a lot of them to count.

A quicker method is to count all the squares in a column at once, by calculating the area of a rectangle or a trapezoid.

DEFINITE INTEGRALS BY RECTANGLES

This sketch has six rectangular columns that can be used to estimate the definite integral.

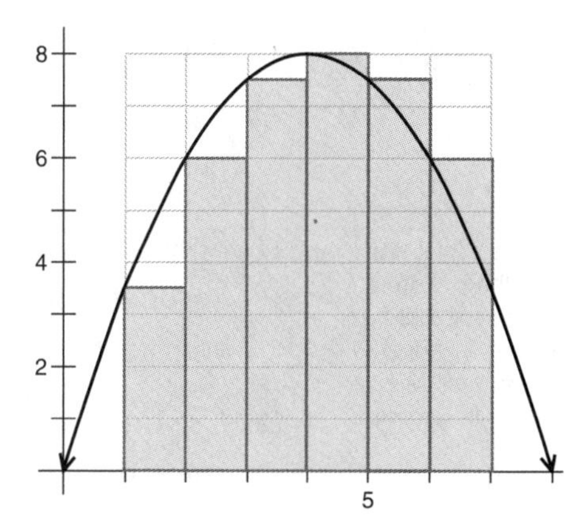

1. Open the sketch **Trapezoidal Accumulation.gsp** from the **9 Introduction to Calculus** folder.

Q1 What is the sum of the areas of the rectangles?

Q2 Do you think this sum is a good estimate of the definite integral? Why or why not?

2. Press the *Animate Grid* button to change the number of rectangles used in estimating the definite integral.

Q3 Watch the sum measurement as the animation progresses. What is the largest estimate for the definite integral? What is the smallest?

3. Stop the animation, and choose **Undo** from the Edit menu to return the number of rectangles to 6.

4. Select two measurements: *width* and *Sum of Rectangle Areas.* Choose **Tabulate** from the Graph menu to create a table.

5. With the table selected, choose **Add Table Data** from the Graph menu. Click the radio button to add 10 entries as values change, and then click OK.

6. Press the *Animate Grid* button again. This time the various estimates are recorded in the table.

Q4 Which of the estimates recorded in the table do you think is the most accurate? Why do you think so?

CONSTRUCT YOUR OWN RECTANGLES

7. Go to page 2 of the sketch. On this page you will construct some rectangles of your own.

8. Press and hold the **Custom** tools icon. From the menu that appears, choose the **Rectangle** tool.

9. You'll use the tool to construct three rectangles. First, click on the x-axis at the start of the domain, exactly where $x = 1.0$. This should produce a single rectangle, along with an area measurement.

If you haven't done it correctly, choose **Undo** from the Edit menu as many times as necessary, choose the **Rectangle** tool again, and then try again from step 9.

10. Click on the bottom right corner of the rectangle you just constructed to make a second rectangle. Click on the bottom right corner of the second rectangle to make the third. If you've constructed the rectangles correctly, the last rectangle will end exactly at $x = 7.0$.

Q5 Each time you used the **Rectangle** tool, a measurement appeared that showed the area of that rectangle. What is the sum of the three areas?

You can use **Parameter Properties** to determine the amount by which the parameter will change when you press the + or – key on the keyboard.

11. Select both the width and height measurements, and press the – key on the keyboard twice. This will change both values from 2.0 to 1.0, so the widths of your rectangles are now 1.0. Now you need more rectangles.

12. Use the **Rectangle** tool again to make the three additional rectangles required to cover the entire domain to $x = 7.0$.

Q6 Now that you have six rectangles, use Sketchpad's Calculator to find the sum of their areas. Is this more or less accurate than your previous result?

DEFINITE INTEGRALS BY TRAPEZOIDS

Rectangles don't do a very good job of estimating the squares that are only partially within the area we want to add up. Using trapezoids will give a more accurate result.

13. Go to page 3 of the sketch. Choose the **Trapezoid** custom tool.

Clicking with the **Trapezoid** tool will produce a single trapezoid, along with an area measurement.

14. Click on the x-axis at the start of the domain, exactly where $x = 1.0$.

15. Click repeatedly along the x-axis to construct five more trapezoids, ending at $x = 7.0$.

16. Add up the six area measurements.

Q7 Use the Calculator to add up the areas of the trapezoids. What is the sum?

Q8 How does this area compare with the area of the rectangles from the previous page? Which do you think is more accurate?

The more accurately you measure the area, the closer you come to the exact value of the definite integral. You can define the definite integral as the limit of the total area as the width of the approximating rectangles or trapezoids decreases.

Q9 If you use 12 trapezoids each of width 0.5, how much does the total area change?

Limits with Tables

The definition of both derivative and integral involves the concept of a limit. In this activity you'll explore what happens when you evaluate a function at values closer and closer to a particular given value. This process is known mathematically as taking the limit of the function. You'll do so by making a table of values and choosing values closer and closer to the x-value at which you want to find the limit.

A TABLE OF VALUES

1. Open the sketch **Limits By Table.gsp** from the **9 Introduction to Calculus** folder. Note that this sketch contains the function $f(x) = \frac{0.4x^2 - 10}{x - 5}$.

2. Use Sketchpad's Calculator to evaluate $f(5)$.

calculate $f(5)$, choose **easure | Calculate,** en click on the function x) in the sketch, press 5 the Calculator's ypad, and click OK.

Q1 What is the result? Why do you think you get this result?

3. The sketch contains a parameter x. Double-click this parameter to change its value. First change it to 4, and record the value of the function $f(x)$. Then change x to 5, and record the function's value. Finally, change it to 6, and record it again.

Q2 Find: $f(4)$ = __________ $f(5)$ = __________ $f(6)$ = __________

4. Change the value of the parameter x to 4.5, and press the *Animate by 0.1* button.

Q3 What happens? What values does x take on? When does it stop?

5. You can collect the changing values in the table on the right side. Select the table and choose **Add Table Data** from the Graph menu. Click the radio button to add 10 entries as the values change, and click OK.

ake sure the value x is reset to exactly 5 before collecting le data.

6. To actually collect the values in the table, the values must be changing. Press the *Animate by 0.1* button to change the parameter and add values to the table.

Q4 What does the table show for $f(5)$? Can you see a pattern in the values of $f(x)$ before and after $f(5)$? If you follow that pattern, what is the value for $f(5)$? (This value is the limit of $f(x)$ as x approaches 5.)

GET CLOSER TO THE LIMIT

7. To get values of $f(x)$ that are closer to this limit, you need to use values of x that are closer to 5. Double-click the x parameter, and change its value to 4.95. Then add 10 more entries to the table, and press the *Animate by 0.01* button to actually add entries.

our table grows too ge, create a second le, or select the sting table and oose **Graph | Remove ble Data.**

Q5 Does the pattern still indicate the same limit as x approaches 5? How close do the new values actually come to the limit?

You may have to drag the table vertically to see the new entries.

8. To get values even closer to the limit, change x to 4.995 and add 10 more entries, using the *Animate by 0.001* button.

x	f(x)
4.95000	3.98000
4.96000	3.98400
4.97000	3.98800
4.98000	3.99200
4.99000	3.99600
5.00000	undefined
5.01000	4.00400
5.02000	4.00800
5.03000	4.01200
5.04000	4.01600

Q6 Does the pattern still indicate the same limit as x approaches 5? How close do these new values actually come to the limit?

Consult your table to answer the next questions.

Q7 Approximately how close must x be to 5 so that the value of $f(x)$ is within 0.01 of the limit?

Q8 Approximately how close must x be to 5 so that the value of $f(x)$ is within 0.001 of the limit?

EXPLORE MORE

Factor the numerator of the original function. Does this shed any light on its behavior

If you need to remove entries from a table, select the table and choose **Graph | Remove Table Data.**

Change the function to one of the following, and collect similar data. For each function, describe your findings, state whether or not the function has a limit L, and give a value of L (if possible).

$$f(x) = \frac{1}{x-5} \text{ at } x = 5$$

$$f(x) = 2 + \frac{|x-5|}{x-5} \text{ at } x = 5$$

To change a button's properties, select the button by clicking on its handle. Then choose **Edit | Properties | Animate.**

The animation buttons in this sketch are set to animate x on both sides of the value 5.0. You can change the properties of the buttons to use a different domain. Modify the buttons to investigate one or more of the following limits:

$$f(x) = |x-2| \text{ at } x = 2$$

$$f(x) = \frac{1}{x^2-1} \text{ at } x = 1$$

$$f(x) = 0.5x + \frac{|x-3|}{x-3} + 1 \text{ at } x = 3$$

$$f(x) = \sin(x) \text{ at } x = \frac{\pi}{4}$$

$$f(x) = \tan(x) \text{ at } x = \frac{\pi}{2}$$

Limits with Delta and Epsilon

The formal definition of a limit is as follows:

> L is the limit of $f(x)$ as x approaches c if and only if
>
> For any $\epsilon > 0$, no matter how small,
> there exists a positive number δ such that,
> when x is within δ of c,
> $f(x)$ is within ϵ of L.

is the Greek letter psilon, and δ is the eek letter *delta*.

In this activity you'll determine the limit L of a function, and you'll adjust the value of δ to satisfy the definition above. For some functions and values of c this will work, because the limit exists. For other functions and values of c this won't work, because there is no such limit.

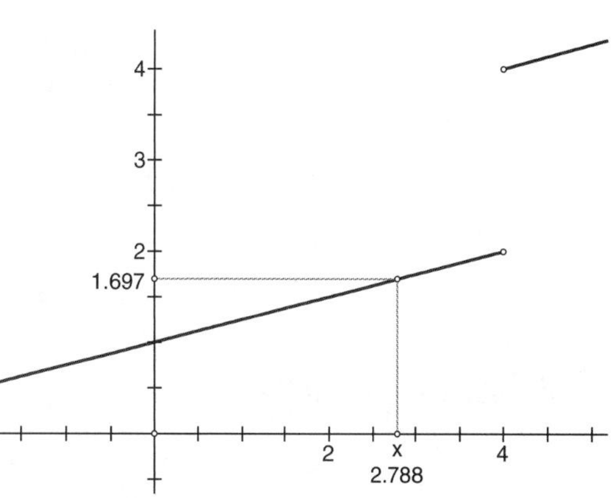

FIRST FUNCTION

1. Open the sketch **Limit Epsilon Delta.gsp** from the **9 Introduction to Calculus** folder. Note that this sketch contains the function $f(x) = \frac{0.4x^2 - 10}{x - 5}$.

2. Double-click the parameter c, and set its value to 5. Drag point x along the axis, and observe the values of x and of the function.

Q1 Use the *Move x ->c* button to move x to the exact value of c. What happens? What is the value of $f(c)$?

3. Drag x back and forth near c, and observe the values of x and $f(x)$.

Q2 What do you think is the limit of the value of the function when x is close to c?

4. Double-click the parameter L, and change its value to the limit from Q2.

5. Press the *Show Epsilon* button to show, in green, a range extending by ϵ above and below L on the y-axis. Use the ϵ slider to change the value of epsilon, and observe the results. Manipulate the slider so that ϵ is approximately 0.25.

confirm numerically, serve the value of $|f(x) - L|$.

Q3 Drag x back and forth. For what values of x is the value of the function within ϵ of L? (In other words, for what values of x does the horizontal red segment touch the green portion of the y-axis?)

6. Press the *Show Delta* button. Drag the δ slider back and forth, and observe the effect on the horizontal blue portion of the x-axis. Press the *Restrict x* button, and drag x back and forth. Observe that the value of x is now restricted to the blue area.

Your job now is to set δ so that the value of $f(x)$ is always within ϵ of L.

Use the *Show Segments* button to help you make this adjustment.

7. Drag the δ slider until you think this condition is met. That is, make the blue portion of the x-axis small enough so that the value of the function will always b within the green portion of the y-axis. Drag x back and forth to test your result.

Q4 What value of δ is required to keep $f(x)$ within ϵ of L?

8. Now try a smaller value of ϵ. Set $\epsilon = 0.15$, and determine the value of δ require to keep $f(x)$ within ϵ of L.

Q5 What value of δ is now required to keep $f(x)$ within ϵ of L?

Q6 Do you think you can find an appropriate value of δ, even if ϵ is very small? Explain your answer.

SECOND FUNCTION

9. Go to page 2 of the sketch. On this page you'll investigate the limit of a different function. Set the value of c to 4.0, and set L to 3.0.

10. Press the *Show Epsilon* button and set ϵ to 1.5. Also press the *Show Delta* and *Show Segments* buttons.

Q7 Can you adjust δ to keep $f(x)$ within ϵ of L? What value of δ must you use?

11. Now choose a smaller value of ϵ, by setting $\epsilon = 0.50$. Try to adjust δ to keep $f(x)$ within ϵ of L.

Q8 What happens when you try to adjust δ to keep $f(x)$ within ϵ of L? Does it help to use a smaller value of δ? If not, try using a different value of L. Can you find a value of L that makes it work?

OTHER FUNCTIONS

12. Go to page 3, containing the function $f(x) = 1 + \frac{x}{x-3}$. Set the value of c to 3.0. Set $\epsilon = 1.00$. Experiment with various settings of L and δ.

Q9 Are there any values of L and δ that keep $f(x)$ within ϵ of L? If so, what values did you use? If not, why not?

Try some of the following functions:

$$f(x) = \frac{1}{x^2 - 1} \text{ at } x = 1 \text{ and at } x = 0$$

$$f(x) = 0.5x + \frac{|x - 3|}{x - 3} + 1 \text{ at } x = 0.5 \text{ and at } x = 3$$

$$f(x) = \tan(x) \text{ at } x = \frac{\pi}{4} \text{ and at } x = \frac{\pi}{2}$$

Manually Probing the Antiderivative

You've already learned about the derivative of a function, and you know that the value of the derivative is the slope of the line that's tangent to the graph of the function.

If $g(x)$ is the derivative of $f(x)$, then $f(x)$ is the antiderivative of $g(x)$.

In this activity, you'll reverse the process and start with the derivative function. Your job is to find the original function for which the given function is the derivative. This original function is called the *antiderivative.* You'll do this by building a probe based on the relationship between a derivative and the slope of a tangent line.

CREATE THE FUNCTION

Use the Values pop-up menu in the Calculator to create parameters a, b, c, and d.

1. In a new sketch, choose **Graph | Plot New Function** and graph the function $f(x) = ax^3 + bx^2 + cx + d$. Use these parameter values: $a = 0.05$, $b = 0.10$, $c = -1$, and $d = -1$. This function is the derivative of the function you want to find.

Q1 For what values of x is the value of the derivative positive? What does a positive derivative tell you about the antiderivative?

BUILD THE PROBE

The probe will be a short segment pointing in the same direction as a tangent line to the antiderivative. The slope of this tangent is equal to the value of the derivative.

Q2 When the derivative is positive, in what direction would you expect the probe to point? In what direction would you expect it to point when the derivative is zero?

2. Construct independent point P, and measure its x- and y-coordinates. This is the starting point of the probe. Calculate $f(x_p)$, the value of the function at P.

Q3 What is the relationship between the value of $f(x_p)$ and the slope of the line that's tangent to the unknown antiderivative function at x_p?

To construct a line through P with the correct slope, you need to find a second point on the tangent line. The first point (point P) has coordinates (x_p, y_p), so you can think of a second point as $(x_p + \Delta x, y_p + \Delta y)$.

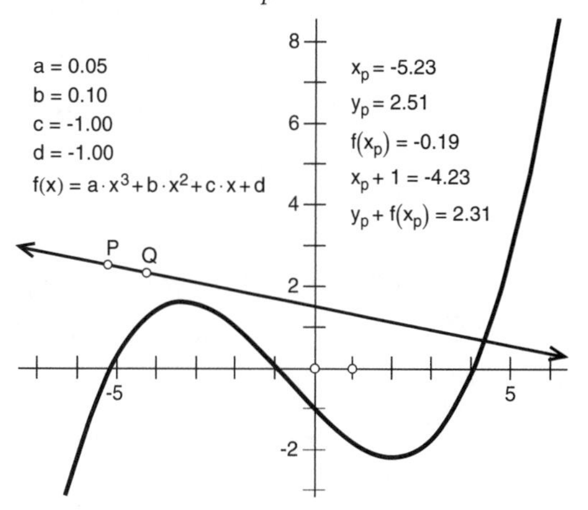

Q4 Remember that you can express the slope of a line as $m = \Delta y/\Delta x$. If you use 1 as the value of Δx, what would be the value of Δy?

3. Calculate the two values $x_p + 1$ and $y_p + f(x_p)$, and plot point Q at $(x_p + 1, y_p + f(x_p))$. Construct a line through points P and Q.

4. Drag point P, and observe how the slope of the line depends on P.

Q5 For what values of x does the line point up to the right? For what values of x is it horizontal? For what values does it point down to the right? Explain these observations based on the behavior of the function $f(x)$.

Although this line points in the correct direction, it's too long to make a good probe. A probe should be a short segment starting from point P.

Now you'll use the line to construct a short probe.

5. Create a parameter $h = 0.5$ to determine the length of the probe segment.

6. Plot point R at $\left(x_p + h, y_p\right)$ to create a point exactly 0.5 away from point P.

7. Construct a circle centered at P and passing through R, and construct the right-most intersection S of the circle with the tangent line. Then hide the tangent line, the circle, and points Q and R.

8. Finally, construct segment PS to be your probe. Drag point P around again, and observe the behavior of the probe.

USE THE PROBE

Because point P snaps to integer values, the segment's traces appear at integer values on the grid.

9. Turn on tracing for the segment, and choose **Graph | Snap Points.** Drag P all around the screen, and observe the traces left behind.

Q6 What do the traces indicate about the function whose derivative is $f(x)$?

Now you'll use the probe to trace a solution to the original problem.

10. Turn off **Snap Points,** drag P so that its x-coordinate is approximately -7, and erase the existing traces.

11. Drag point P so that it follows the direction of the probe. In other words, if the probe is pointing up and to the right, drag P up and to the right, trying to follow the probe segment. As you drag, the slope of the probe will change; as the slope changes, continue dragging P in the direction that the probe points.

12. Practice following the probe several times, erasing traces before each new attempt. See how good you can get at following the traces accurately.

This second trace should start from the same x-value as the original, but start at a different y-value.

13. Once you have a good smooth trace, leave it on the screen, and quickly move the probe so that it's directly above or below the starting point for the existing trace. Without erasing the original trace, make another trace.

14. Make a third trace, starting at the same x-position but a different y-position.

Q7 What do you notice about the shapes of the three traces?

15. Save your finished sketch. You'll need it for the next activity.

utomatically Probing the Antiderivative

In the last activity you created a probe and used it to trace the antiderivative of a function. In this activity you'll automate the process of tracing the antiderivative. You'll do this in two ways, using a movement button and using the **Iterate** command.

Open the sketch you saved in the last activity. This sketch contains the function $f(x)$ and a probe from point P to point S whose slope matches the value of $f(x_p)$.

RACE WITH A MOVEMENT BUTTON

1. Select points *P* and *S*, in that order, and choose **Edit | Action Button | Movement.** Accept the default settings for the movement button.

ake sure that tracing active for the probe. oose **Display | Trace gment** to turn tracing or off.

2. Drag point *P* to an appropriate starting *x*-value. Press the *Move P->S* button, and point *P* will automatically move in the direction of *S*. As the slope of the probe changes, point *P* continues to move toward the new position of *S* so that the movement button creates a trace automatically.

oint *P* has gone -screen, you can t it back by choosing **it | Undo Translate int.**

3. Once your trace has gone far enough to the right, click the button again to stop the movement.

Q1 How does this automatic trace compare to the manual traces you did previously?

4. Erase the existing traces, and create five new traces of the antiderivative, each one starting from the same *x*-coordinate but a different *y*-coordinate.

Q2 What do you notice about the shapes of the five traces?

RACE BY ITERATION

The traces you made manually and with the *Move P->S* button are not dynamic objects; once you make such a trace, you can't do anything with it except to erase it and make a new one. In the next few steps, you'll use iteration to create a permanent trace that enables you to move the starting point around.

5. Turn off tracing for the probe segment. Don't erase the existing traces just yet.
6. Position point *P* at the same *x*-value that you've been using.
7. With point *P* selected, choose **Transform | Iterate.** In the Iteration dialog box, designate point *S* as the image point by clicking on point *S* in the sketch.

e iteration consists of rated images of both nt *P* and the segment nnecting *P* and *S*.

8. Press Iterate in the dialog box to iterate the probe three times. The original probe (the pre-image) goes from *P* to *S*; the first image is constructed using the same method, but beginning at point *S*; the second image starts at the ending point of the first image, and so on.

9. With the iterated image selected, press the + key on your keyboard several time to increase the number of iterations.

Q3 How does the shape of the iteration compare with the existing traces on the screen?

The iteration may not look as smooth as the traces, because each step of the iteratio is a segment that's longer than the steps point *P* took when you pressed the action button. Fortunately, you can reduce the size of the segments in the iteration.

10. Change the value of h to 0.1.

Q4 Describe what happens to the iteration. (You should note at least two importan changes.) Explain why changing the value of h has this effect.

11. To increase the number of steps in the iteration, you can select the iteration and press the + key on the keyboard, but with this method it will take you some tim to create 1000 iterations. Select the iterated image of either the point or the segment and choose **Edit | Properties.** Use the Iteration panel to change the number of iterations to 1000.

12. Change the value of h to 0.01.

Q5 How does this new value change the shape of the iteration? Do you think this shape is a more accurate depiction of the antiderivative? Why?

Q6 This improved iteration is still an approximation. What would you have to do tc make the shape correct to any desired degree of accuracy?

13. Move point *P* vertically on the screen so that the iteration starts at a new y-value.

Q7 How does the shape of the antiderivative change as you move *P* up and down?

Q8 For each different position of point *P*, there's a different antiderivative. How many solutions are there to the question "Find a function whose derivative is $f(x)$?" Explain your answer.

EXPLORE MORE

Q9 Change the original function $f(x)$ by changing the values of the parameters a, b, c, and d, and note how the shape of the antiderivative relates to the changed shape of the function itself.

Q10 Change the original function to $f(x) = a\sin(bx + c) + d$. What is the resulting shape of the antiderivative? Try different values of a, b, c, and d to experiment with different sinusoids.

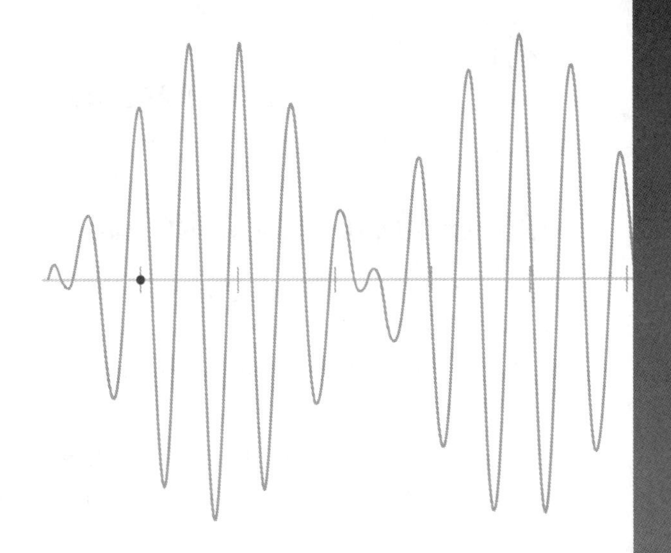

Activity Notes

TRANSLATION OF FUNCTIONS (PAGE 3)

Objective: Students create both geometric and algebraic translations of a function, match the two translations, and use the components of the geometric vector to write an equation for a translated image.

Prerequisites: Students should be familiar with geometric translations, translation vectors, graphs of functions, and coordinates.

Sketchpad Proficiency: Beginner. This activity contains detailed instructions to guide students who have never used Sketchpad before.

Class Time: 25–30 minutes for students already experienced with Sketchpad, or 40–45 minutes for students being introduced to Sketchpad.

Required Sketch: Translate.gsp

Presentation Sketch: Translate Work.gsp

EXPLORE TRANSLATIONS OF FUNCTIONS

In this activity, students relate two different representations of the translation of a function. They begin by constructing a geometric translation of a function plot, and then construct an algebraic translation of the same function. Encouraging students to make connections between geometric and symbolic representations of functions is a primary goal of this activity.

Following the completion of this activity, you should have a class discussion of the distinctions and connections between the two representations. (In Sketchpad, there's a related distinction between the function object—in symbolic form—and the function plot—in geometric form.)

For many students this first activity will be their introduction to Sketchpad. To accommodate both experienced and beginning Sketchpad users, the first two section's of the activity have two subsections: a Summary and Detailed Instructions. The experienced user can read the Summary, do what it calls for, and skip the Detailed Instructions. The beginner can read the Summary as a high-level overview before following the Detailed Instructions to perform the steps that the Summary describes.

This activity requires only the **Arrow** tool and the **Text** tool. Although there are tips about using these tools in the margin notes, students new to Sketchpad will benefit from a brief demonstration of how to select and deselect objects using the **Arrow** tool, and how to show, hide, and change labels using the **Text** tool.

Q1 The purpose of the first question is to introduce students to the idea of translating functions and to encourage them to think about translations before they start constructing. The actual answer that students provide is not as important as the process of speculating about possible outcomes of translation.

TRANSLATE A POINT ON THE FUNCTION PLOT

In this section, students prepare for the geometric translation of the function by translating a single point on the original plot. The locus of this translated point will be the geometric translation of the function plot.

TRANSFORM THE FUNCTION PLOT GEOMETRICALLY AND ALGEBRAICALLY

Q2 It is important that students answer this question before they construct the locus of P'. Otherwise, they may not understand the result of the **Locus** command.

In this section, students will create both geometric and algebraic translations, manipulate the first by changing the vector, and manipulate the second by changing sliders.

7. Though students don't construct anything new in this step, it's very important that they pay attention to the effects of dragging point P and point V. In this and other steps, understanding why dragging particular objects produces the effects that it does is a critical part of the discovery process.

MATCH THE TRANSFORMATIONS

In this section, students manipulate the two translations so that they match. Comparing the slider values and the coordinates of the translation vector will lead students to the general equation for the translated image of a function.

Q3 and Q4 The image always has the same shape, size, and orientation as the pre-image. The purpose of these questions is to emphasize the characteristics of translation.

Q5 The graph moves left if $a > 0$ and right if $a < 0$. Explanations will vary. One explanation is that the translated function takes the same value at $x = 0$ as the original function did at $x = a$. If $a > 0$, the point on the image that corresponds to $x = a$ on the original will have moved a units to the left, to $x = 0$.

Q6 The motion of the function plot is up for $b > 0$ and down for $b < 0$. As in Q5, explanations vary.

Q7 Because $b = 0$, the algebraic translation is horizontal and vector OV must be horizontal to make the geometric translation match. Values of x_V and y_V are shown here.

	a	b	x_V	y_V
Q7	3	0	−3	0
Q8	0	−2	0	−2

Q8 Because $a = 0$, the algebraic translation is vertical and vector OV must be vertical.

CONCLUSION

Q9 and Q10 Students will add two more rows to the table above to confirm their observations. Recording these results and comparing slider values and vector components will help students to generalize their observations.

Q11 Students should conclude that $a = -x_V$ and $b = y_V$, though different students may express this conclusion differently.

13. The final equation for the translated function is

$$g(x) = f(x - x_V) + y_V$$

Q12 If you edit the equation as in step 13 above, the geometrically translated function and the algebraically translated function will always match.

EXPLORE MORE

Q13 By trying several functions, students should conclude that the equation entered in step 13 is general and does not depend on the nature of $f(x)$.

Q14 Students are challenged with the question of uniqueness of a function translation. In order to explore this equation, students need to go to the Challenge page.

The focus of the challenge is to find a function that can be translated in more than one way to produce the same image. See the file **Translate Challenge Notes.pdf** on the CD-ROM for teacher notes on the challenge problem.

DEMONSTRATE

The sketch **Translate Work.gsp** contains completed constructions for the activity. You can use it for presentation to the class. The hide/show buttons allow you to pose questions first and then show the results of the constructions that follow the student activity. Pages 1–4 correspond to the main activity and pages 5–8 correspond to Q14 of the Explore More section of the activity.

DILATION OF FUNCTIONS (PAGE 8)

Objective: Students explore dilations of functions geometrically and algebraically and learn to write equations for dilated functions.

Prerequisites: Students should be familiar with geometric dilations and scale factors.

Sketchpad Proficiency: Beginner to Intermediate, but contains detailed instructions for performing the various steps. Students who have done the preceding activity, Translation of Functions, will be able to do this activity.

Class Time: 30–40 minutes

Required Sketch: Dilation.gsp

Presentation Sketch: Dilation Present.gsp

EXPLORE STRETCHING AND COMPRESSING

Various textbooks use different terminology for this topic. The title of the activity uses *dilation* in order to emphasize the connection between geometry and algebra. *Dilation* (as it is used in geometry) also has the advantage of referring to images that can be either larger or smaller than the pre-image.

Q1 This image has been stretched horizontally by a factor of two and compressed vertically by a factor of one-half.

TRANSFORM THE FUNCTION PLOT GEOMETRICALLY AND ALGEBRAICALLY

Q2 The path of the plotted point is similar to the graph of $f(x)$, stretched horizontally by a factor of two.

Q3 The path is similar to the graph of $f(x)$ compressed vertically by a factor of one-half.

Q4 This time the original graph is compressed horizontally by a factor of one-half and stretched vertically by a factor of two.

MATCH GEOMETRIC AND ALGEBRAIC TRANSFORMATIONS

Q5 When $d = 1$, the graph of $g(x)$ is stretched vertically if $c > 1$ and compressed vertically if $c < 1$.

Q6 When $c = 1$, the graph of $g(x)$ is compressed horizontally if $d > 1$ and stretched horizontally if $d < 1$.

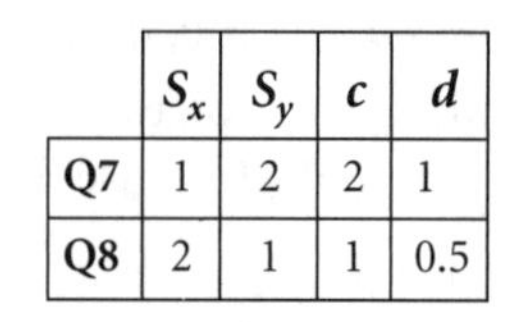

	S_x	S_y	c	d
Q7	1	2	2	1
Q8	2	1	1	0.5

Q9 and Q10 Answers will vary, but all answers fit the pattern described in Q11.

Q11 In every case, $c = S_y$ and $d = 1/S_x$.

13. The formula is $g(x) = S_y f(x/S_x)$.

Q12 This formula for $g(x)$ will work no matter what function $f(x)$ represents. The specific functions students try for $f(x)$ will vary.

During discussion after the activity, you should ask students whether they experimented with negative scale factors; if they did, ask them to describe the result.

EXPLORE MORE

This section gives students a chance to combine what they've learned in this activity with what they learned in the Translation of Functions activity. Each challenge requires both translation and dilation to solve.

Q13 The equation is

$$h(x) = S_y \cdot f\left(\frac{x - x_v}{S_x}\right) + y_v$$

Q14 The equations are different, because it makes a difference whether you translate or scale first. The equation is

$$h(x) = S_y\left(f\left(\frac{x}{S_x} - x_v\right) + y_v\right)$$

Q15 Answers will vary but must conform to the patterns described in Q16.

Q16 In Challenge 1, the dilation requires a combination of S_x and S_y such that $2S_y = S_x^2$. The translation will vary according to the values of S_x and S_y. You can make the shapes of any two parabolas equivalent by either a horizontal or a vertical dilation. In Challenge 2, the dilation requires that $S_y = \pm 1$ and $S_x = \pm\frac{1}{4}$, and the translation requires that $y_v = -2$. If S_x and S_y have the same sign, the value of x_v can be any multiple of $\pi/2$, because you can transform any sinusoidal function into itself by a horizontal translation. (If S_x and S_y have different signs, you must offset the value of x_v by $\pi/4$.) In Challenge 3, $y_v = 0$ and $S_x = 1$, but various combinations of x_v and S_y such that $x_v = \log_2 S_y - 1$ give the desired transformation. This equivalence is due to the relationship between multiplication of numbers and addition of exponents.

Consider asking students who undertake this question to show how the different sets of slider values—that is, the different equations for $h(x)$—are equivalent algebraically.

Q17 You can translate linear functions in a variety of ways and dilate arbitrarily.

Q18 Assuming only positive values (that is, assuming S remains in the first quadrant), the image function is stretched in the x-direction as you drag point S to the right and will be compressed as you drag S to the left. Similarly, the image is stretched in the y-direction as you drag S up and compressed as you drag S down.

DEMONSTRATE

The document **Dilation present.gsp** has completed constructions for the activity. You can use it for a presentation to the class. On pages 1 and 2 there is an animation button for the point P, so you can easily trace the image function.

REFLECTION OF FUNCTIONS (PAGE 12)

Objective: Students explore reflections across the x- and y-axes algebraically and geometrically and write equations for reflected functions.

Prerequisites: Students should be familiar with geometric reflections and dilations.

Sketchpad Proficiency: Intermediate. Students turn tracing on and off, reflect points across mirrors, merge points, and plot and edit functions.

Class Time: 20–30 minutes

Required Sketch: Reflection.gsp

Presentation Sketch: Reflection Present.gsp

INVESTIGATE REFLECTIONS

Students will explore properties of reflections across the x- and y-axes. So that students can see the results of these transformations, they work with a limited-domain function confined to the first quadrant.

Q1 The plot of g is the mirror image of the plot of f over the y-axis, and the plot of h is the mirror image of the plot of f over the x-axis.

Q2 The trace of Q is a mirror image of the trace of P over the y-axis, and the trace of R is the mirror image of the trace of P over the x-axis.

Q3 $x_Q = -x_P$; $y_Q = y_P$; $x_R = x_P$; $y_R = -y_P$

Q4 Point Q: $g(x) = 1 + \sqrt{-x - 1}$

Point R: $h(x) = -1 - \sqrt{x - 1}$

Q5 Multiplying the value of the function by -1 produces $h(x) = -f(x)$, which corresponds to the reflection across the x-axis. Multiplying the value of x by -1 produces $g(x) = f(-x)$, which corresponds to the reflection across the y-axis.

Q6 Two reflections, one over each coordinate axis, match the third function. Alternative answers include a 180° rotation about the origin and a dilation by -1.

Q7 The relationship is true for all functions.

EXPLORE MORE

In this section, students investigate reflections across arbitrary horizontal and vertical lines. They also make connections between reflection and dilation. This provides students an opportunity to combine what they have learned about translations, dilations, and reflections.

Q8 Reflection across the line $x = a$ is equivalent to reflection over the y-axis and translation by $2a$: $g(x) = f(-x + 2a)$. Reflection across the line $y = b$ is equivalent to reflection over the x-axis and translation by $2b$: $h(x) = -f(x) + 2b$.

Q9 Reflection across the y-axis is equivalent to horizontal dilation by -1. Reflection across the x-axis is equivalent to vertical dilation by -1.

DEMONSTRATE

The document **Reflection Present.gsp** has completed constructions for the activity. You can use it for a presentation to the class. Each page of the document has action buttons that provide you with an opportunity to ask questions and demonstrate the results.

ABSOLUTE VALUE OF FUNCTIONS (PAGE 14)

Objective: Students explore absolute value transformations and investigate the connection between absolute value and reflection.

Prerequisites: Students should be familiar with reflections of functions.

Sketchpad Proficiency: Intermediate

Class Time: 20–25 minutes

Required Sketch: Absolute Value.gsp

Presentation Sketch: Absolute Value Present.gsp

EXPLORE ABSOLUTE VALUES

The goal of this exploration is for students to understand the effect of absolute value operations on the plot of a function and to make connections between the absolute value and geometric reflection.

Q1 and Q2 Points P and Q trace g. When $y_P \geq 0$, P traces g, and when $y_P < 0$, Q traces g.

Q3 The value of y_P represents the output of function f. Function g takes the absolute value of the output of f, so the value of g is always positive. When y_P is positive, it represents the correct output from g, so P is on g. But when y_P is negative, the corresponding positive value of y is represented by the reflection across the axis (point Q), so in this portion of the graph Q is on g.

Q4 Function h uses the absolute value of the input of f, so function h corresponds only to that part of f to the right of the y-axis. When x_P is positive, the current output of the function appears for both positive x_P and for the value of $-x_P$, corresponding to the reflection across the y-axis (point R). Therefore, point R traces out the left portion of the graph.

Q5 When $x_P \geq 0$ and $y_P \geq 0$, points P and R (the y-axis reflection) trace the right and left portions of j, just as they did with function h. But when $x_P \geq 0$ and $y_P < 0$, the output of j must be positive, even though y_P is negative. For this part of the graph, points Q and

S (the x-axis reflections) travel along the right and left portions of j. When $x_P < 0$, none of the four points lies on j.

Q6 The functions investigated will vary, but the conclusions are the same no matter what function is used. Here's the filled-in table:

point(s)	trace(s) function	when	and point(s)	trace(s) function	when
P	f	$x_P \geq 0$	P	f	$x_P < 0$
P	g	$y_P \geq 0$	Q	g	$y_P < 0$
P	h	$x_P \geq 0$	R	h	$x_P \geq 0$
P & R	j	$x_P \geq 0$ $y_P \geq 0$	Q & S	j	$x_P \geq 0$ $y_P < 0$

EXPLORE MORE

This section challenges students' understanding of the effect of different transformations on the plot of a function. Students can plot test functions using translation vector and scaling factors. They can also use the two custom tools for the absolute value transformation.

Challenge 1: $g(x) = -|f(-x + 1)|$

Challenge 2: $g(x) = -f\left(\frac{|x|}{2}\right) + 2$

DEMONSTRATE

The document **Absolute Value Present.gsp** has completed constructions for the activity. You can use it for a presentation to the class.

COMPOSITION OF FUNCTIONS (PAGE 16)

Objective: Students use dynagraphs to analyze composition of functions, domain and range of composite functions, and commutativity of function composition.

Prerequisites: Students should be familiar with domain and range and with the commutative property.

Sketchpad Proficiency: Beginner to Intermediate. The activity relies on a prepared sketch in which students must split and merge points and edit function definitions.

Class Time: 30–40 minutes

Required Sketch: Composition.gsp

Presentation Sketch: Composition Present.gsp

EXPLORE COMPOSITE FUNCTIONS

The functions presented on page 1 of the sketch are intended as an example only and are not intended as an actual model of the complicated physical processes of global warming and melting of ice caps. Students interested in this topic may want to read the detailed article in the March 2004 *Scientific American,* available at www.sciam.com/ontheweb.

Q1 The input for the second dynagraph must be equal to the output of the first dynagraph.

Q2 The linked dynagraphs show $g(f(x))$, since $x = A$ is an input for $f(x)$ and then $B = f(A)$ is an input for the function $g(x)$. Thus, the final output is $g(B) = g(f(A))$.

Q3 The value of h exists only for $-4 \leq x \leq 7$.

Q4 $y = h(g(f(x))$ or $y = (h \circ g \circ f)(x)$.

Q5 Answers vary.

DOMAIN, RANGE, AND COMMUTATIVITY OF COMPOSITE FUNCTIONS

Q6 and Q7 Domain $[2, 2.5)$ and $(2.5, 4.5]$, range: $(-\infty, -1]$ and $[0.25, \infty)$.

Q8 First pair of functions: domain $[4, 7]$, range $[0, 6]$. Second pair of functions: domain $[0, 4]$, range $[3, 12]$.

Q9 Conjectures will vary.

Q10 Function composition is not commutative in the general case. For the same value of x, $f(g(x)) \neq g(f(x))$ for the functions given.

EXPLORE MORE

In this section, students explore special cases when $f(g(x)) = g(f(x))$. This section serves as a natural link to the concept of inverse functions. Students also explore the graphs of composite functions on the coordinate axes.

Q11 $f \circ g = 0.5(x - 2) \neq g \circ f = 0.5x - 2$

$f \circ g = (x^3)^2 = g \circ f = (x^2)^3$

$g \circ f = \sqrt{x^2} = |x| \neq f \circ g = (\sqrt{x})^2 = x, x \geq 0$

Because of the square root, the domain of g is restricted, so the composite functions differ: they have different domains.

$f \circ g = g \circ f = x$. This is the case of inverse functions.

Q12 The functions must have the same domain. Unless the functions are inverses, they must be analyzed on a case-by-case basis.

Q13 The plots of $u(x)$ and $v(x)$ are different.

Q14 Students can plot a point with coordinates $(f(A), 0)$ or use the identity line to determine a point on the x-axis with $x = f(A)$. See **Composition Present.gsp** for a completed construction.

DEMONSTRATE

The document **Composition Present.gsp** has completed constructions for the activity. It can be used for a presentation to the class.

INVERSES OF FUNCTIONS (PAGE 19)

Objective: Students explore inverses of functions graphically and symbolically and determine when inverse functions exist.

Prerequisites: Students should be familiar with function composition, reflection symmetry, and the vertical line test.

Sketchpad Proficiency: Intermediate. Students plot and edit functions, measure coordinates, plot points, turn tracing on and off, and construct a segment, a midpoint, and a locus.

Class Time: 30–35 minutes

Required Sketch: Inverse.gsp

Presentation Sketch: Inverse Present.gsp

EXPLORE INVERSE RELATIONS

1. Note the dashed vertical segment in the sketch. Students can use this segment to help them determine when $g(f(x)) = x$.

Q1 $g(f(x)) = x$ only when $x = 0.6$, so g is not the inverse of f.

Q2 Case 2: $g(f(x)) = x$ for all x, so g is the inverse of f.

Q3 Case 3: $g(f(x)) = x$ only when $x = 0$, so g is not the inverse of f.

Case 4: $g(f(x)) = x$ for all $x \neq 0$, so g is the inverse of f on the restricted domain $x \neq 0$ only.

Case 5: $g(f(x)) = x$ only for $x \geq 0$, so g is the inverse of f on the restricted domain only.

Case 6: $g(f(x)) = x$ for $-\pi/2 \leq x \leq \pi/2$, so g is the inverse of f on the restricted domain only.

Q4 g is not the inverse of f. The inverse is $g(x) = 2x$.

Q5 $f(x) = 2x - 1$: $g(x) = (x + 1)/2$

$f(x) = x^2$: The inverse does not exist unless the domain is restricted. If f is defined only on $x \geq 0$, $g(x) = \sqrt{x}$, and if f is defined only on $x \leq 0$, $g(x) = -\sqrt{x}$.

$f(x) = x^3$: $g(x) = \sqrt[3]{x}$

Q6 The domain of g is the range of f, and the domain of f is the range of g.

Q7 Switch x and y in the function f and solve for y to get the function g.

EXPLORE PLOTS OF INVERSE FUNCTIONS GEOMETRICALLY

Q8 $g(y_P) = x_P$. The ordered pair (y_P, x_P) belongs to g.

Q9 and Q10 The equation is $g(x) = 2x + 2$. You can determine this function by interchanging x and y in $f(x)$ to obtain $x = 0.5y - 1$ and then solving for y.

Q11 The slope of segment PR is $\frac{y_P - y_R}{x_P - x_R} = \frac{y_P - x_P}{x_P - y_P} = -1$

Q12 The coordinates of point M are

$$\left(\frac{x_P + y_P}{2}, \frac{y_P + x_P}{2}\right)$$

so the locus of M includes all points for which $y = x$.

Q13 The line $y = x$ is the perpendicular bisector of segment PR and is thus a line of reflection for this segment.

Q14 The plot of g is the reflected image of the plot of f across the line $y = x$.

CONCLUSION

Q16 and Q17 Function f must be one-to-one to have an inverse. The vertical line test for the inverse is equivalent to a horizontal line test for the original function. (A one-to-one function passes both the vertical line test and the horizontal line test.)

EXPLORE MORE

On their own, students may determine that the function $y = \frac{a}{x}$ is its own inverse for all real values of a. This is a special case of the rational function result in Explore 2.

Explore 1: A linear function $y = mx + b$ is its own inverse when $m = -1$ or when $m = 1$ and $b = 0$.

Explore 2: Any rational function $y = \frac{ax + b}{cx + d}$ is its own inverse when $a = -d$. Solving for x and switching x and y, we get $y = \frac{-xd + b}{cx - a}$, from which it follows that $a = -d$, and that b and c can be any real values.

DEMONSTRATE

The document **Inverse Present.gsp** contains completed constructions, including action buttons to show various steps, and can be used for a classroom demonstration.

TRANSFORMATION CHALLENGE (PAGE 22)

Objective: Students challenge each other with transformed functions, looking at the graph representation and trying to deduce the algebraic representation.

Prerequisites: Students should be familiar with function translations, dilations, reflections, and absolute values.

Sketchpad Proficiency: Beginner. Students edit function definitions.

Class Time: Variable, but at least 15–20 minutes

Required Sketch: Transformation Challenge.gsp

THE GAME

Students could play this game after any of the activities in Chapter 1. It can add some fun and challenge to the process as students get better at understanding function transformations and at connecting the algebraic and graphical representations of functions.

INTRODUCTION TO RADIANS (PAGE 25)

Objective: Students learn about radian measure and investigate a Sketchpad simulation that approximates π to two decimal places.

Prerequisites: None

Sketchpad Proficiency: Beginner. Students use a pre-made sketch.

Class Time: 20–30 minutes

Required Sketch: Radians.gsp

SKETCH AND INVESTIGATE

Q1 Three complete petals and the start of a fourth petal are formed when point B travels around the circumference.

Q2 For each petal formed, point B travels a distance of $2r$. Note that the length of the petal itself, however, is not $2r$.

Q3 Point B travels a little more than $3 \cdot 2r = 6r$ as it traces the circle's circumference.

Q4 Because π is a little more than 3, the circumference formula $C = 2\pi r$ says that approximately 6 radii fit into a circle's circumference.

Q5 Point A traces 22 petals as point B travels approximately 7 times around the circumference.

From the animation alone, it's hard to tell that 22 petals do not correspond precisely to 7 trips around the circumference. This inexactness can remain unaddressed until Q8, where students obtain $\pi = 22/7$, which is not a precise value of π.

Q6 44 radii are contained in 7 circumferences (again, this answer is approximate).

Q7 Written in equation form, $44r = 7 \cdot 2\pi r$. Solving for π yields 22/7.

Q8 The fraction 22/7 approximates π accurately to two decimal places. For such a simple simulation, that's pretty good!

TRIGONOMETRY TRACERS (PAGE 27)

Objective: Students construct a unit circle and use a point on the circle to define and plot circular functions.

You can combine this activity with the related activity, Transformations of Trigonometric Functions, later in the chapter.

Prerequisites: None

Sketchpad Proficiency: Beginner. Students construct the sketch from scratch using several tools and menu commands, but are given detailed directions for each step.

Class Time: 20–30 minutes

Required Sketch: None

Presentation Sketch: Trigonometry Tracers.gsp

MEASURE THE POINT

Q1 The moving point's horizontal position depends on how far around the circle point C is, and its vertical position depends on the y-value of C.

Q2 The shape is a sine curve. It starts at the origin and is positive from the origin to $x = \pi$. It's negative from $x = \pi$ to $x = 2\pi$. The maximum value is $+1$, and the minimum value is -1. (Students may use numbers rather than π in describing the behavior.)

MAKE MORE MEASUREMENTS

Q3 This shape is a cosine curve. It starts at 1 on the y-axis and is positive until $x = \pi/2$. It's negative from there to $x = 3\pi/2$ and then is positive until it ends at $x = 2\pi$. The maximum value is $+1$, and the minimum value is -1. (Students may use numbers rather than π in describing the behavior.)

16. In discussing this activity after students finish, this step represents a good opportunity to ask students to describe the similarities and differences between the trace they displayed in step 11 and the locus they constructed in step 16.

Q4 This shape is a tangent curve. It starts from the origin and is positive from $x = 0$ to $x = \pi/2$ and from $x = \pi$ to $x = 3\pi/2$. It's negative from $x = \pi/2$ to $x = \pi$ and from $x = 3\pi/2$ to $x = 2\pi$.

Q5 When C is close to the y-axis, the value of x_C is very small, so the value of the fraction y_C/x_C is very large, causing the point to go far off-screen. The corresponding description of the function is to say that its value increases without limit, or to say that the function is undefined at $x = \pi/2$ and at $x = 3\pi/2$.

EXPLORE MORE

Q6 The first function $(1/y_C)$ is the cosecant function, which is positive from the origin to $x = \pi$ and negative from $x = \pi$ to $x = 2\pi$ and is undefined at $x = 0$ and at $x = \pi$.

The second function $(1/x_C)$ is the secant function, which is positive from $x = 0$ to $x = \pi/2$, negative from there to $x = 3\pi/2$, and positive until it ends at $x = 2\pi$. This function is undefined at $x = \pi/2$ and at $x = 3\pi/2$.

The last function (x_C/y_C) is the cotangent. It is positive from $x = 0$ to $x = \pi/2$ and from $x = \pi$ to $x = 3\pi/2$ and negative from $x = \pi/2$ to $x = \pi$ and from $x = 3\pi/2$ to $x = 2\pi$. It is undefined at $x = 0$, $x = \pi$, and $x = 2\pi$.

DEMONSTRATE

To demonstrate this method of defining the six trig functions, you can use the sample sketch that comes with Sketchpad: **Samples | Sketches | Trigonometry | Trigonometry Tracers.gsp.**

SIX CIRCULAR FUNCTIONS (PAGE 29)

Objective: Students create a diagram containing six segments representing the six circular functions and use it to analyze the behavior of the six functions. The diagram can serve as a convenient mnemonic device to help students remember how the functions behave.

Prerequisites: Students should already be familiar with the definitions of the six trig functions.

Sketchpad Proficiency: Intermediate. Students start with a new sketch and build the diagram using the Compass and Straightedge tools and various commands from the Construct, Measure, and Graph menus.

Class Time: 35–40 minutes

Required Sketch: None

Presentation Sketch: Trigonometry Tracers.gsp

This activity is based on a suggestion from John Kerrigan of West Chester University.

SKETCH AND INVESTIGATE

This activity uses the term *circular functions* because the functions are defined based on the position of P on the unit circle, rather than by reference to a particular triangle. If your students have never encountered this term, describe it as an alternate name for the trig functions. Explain that these are the same six functions, and that the choice of terminology is often based on the method used to define the functions.

6. The *sin* and *cos* segments correspond to the normal geometric definitions in $\triangle OPS$, as shown in the figure on the left. The more common geometric definitions of the tangent and secant (segments BC and OC on the right) are replaced in this diagram by the congruent segments PQ and OQ, respectively.

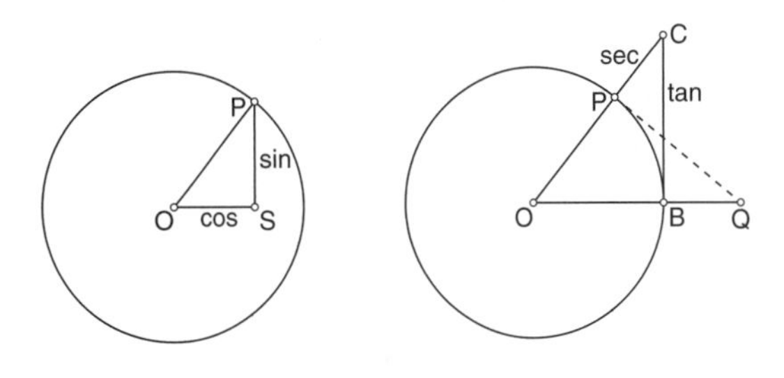

By using these alternate segments, the diagram becomes simpler and easier for students to remember. Another advantage of this arrangement is that the "co-" segments are all on the left, and the other segments (*sin, tan,* and *sec*) are on the right, reducing student confusion between sec and csc.

Q1–Q3 The completed table is shown here.

Segment	Label	QI	→	QII	→	QIII	→	QIV	→	QI
PS	sin	+	1	+	0	−	−1	−	0	+
PT	cos	+	0	−	−1	−	0	+	1	+
PQ	tan	+	∞	−	0	+	∞	−	0	+
PR	cot	+	0	−	∞	+	0	−	∞	+
OQ	sec	+	∞	−	−1	−	∞	+	1	+
OR	csc	+	1	+	∞	−	−1	−	∞	+

13. The value of $\angle BOP$ ranges from $-\pi$ to π. To create a graph from 0 to 2π, use the arc angle from B to P on the circle to measure the angle used for the plotted points.

16 and 17 In these steps, students establish the connection between the geometric segments and the shape of their graphs. This is an important connection that you should emphasize during class discussion of this activity. Once students have mastered this connection, they can re-create the shape of any of the six graphs simply by imagining the segment as point P moves around the circle.

DEMONSTRATE

To present the triangle and relationships in this sketch, use the All in One page of the sample sketch that comes with Sketchpad: **Samples | Sketches | Trigonometry | Trigonometry Tracers.gsp.**

TRANSFORMATIONS OF CIRCULAR FUNCTIONS (PAGE 31)

Objective: Students use a point on the unit circle to define and plot transformed circular functions.

Prerequisites: Students should have completed the activity Trigonometry Tracers earlier in the chapter.

Sketchpad Proficiency: Beginner. Students start with pre-made sketches and are given detailed instructions to complete the construction.

Class Time: 20–30 minutes

Required Sketch: Circular Transforms.gsp

SKETCH AND INVESTIGATE

Q1 $\angle DAF$ is twice as large as $\angle DAC$.

Q2 Point F travels twice around the circle for every revolution of point C.

Q3 When $k = 3$, $\angle DAF$ is three times as large as $\angle DAC$, and F travels three times around the circle for every revolution of C.

Q4 Predictions will vary. The important thing is that students make a prediction.

Q5 The graph is a sine graph compressed in the x direction. It has an amplitude of 1 and a period of $2\pi/3$ so that it shows 3 complete cycles between 0 and 2π.

8. When you change k, the period becomes $2\pi/k$, and the graph shows k complete cycles between 0 and 2π.

Q6 The graphs of these functions resemble the graphs produced in Trigonometry Tracers, but (like the sine plot) compressed in the x direction so that they show k cycles between 0 and 2π.

EXTENSION

You could challenge students to find a way to modify the construction to produce vertical dilation in the resulting graph. One method would be to put a point on segment AF and plot the point's y-coordinate as a function of the angle. If segment AF is constructed as a ray, it's possible to produce both compression and stretching. Alternatively, you could dilate point F toward or away from center point A.

SINE CHALLENGE (PAGE 32)

Objective: Students are challenged to create a sine function so that the first crest of the graph passes through a given point.

Prerequisites: Students should be familiar with the amplitude and period of sine graphs.

Sketchpad Proficiency: Advanced. Students create parameters, modify properties, graph functions, tabulate data, and make a custom tool.

Class Time: 20–30 minutes

Required Sketch: None

Presentation Sketch: Sine Challenge.gsp

SKETCH AND INVESTIGATE

Q1 The value of a always matches the value of y_P.

8. The edited function is $y = y_P \cdot \sin(b \cdot x)$.

9. The period of the function is $2\pi/b$; the value of x_P is 1/4 of a period. Thus $x_P = 1/4 \cdot 2\pi/b$, so $b = \pi/(2x_P)$. The edited function is

$$y = y_P \cdot \sin\left(\left(\frac{\pi}{2x_P}\right)x\right)$$

EXPLORE MORE

The combining of sine functions leads nicely into the activity Sums of Sinusoidal Functions in Chapter 3 and also relates to the activity Taylor Series in Chapter 8.

DEMONSTRATE

Use the sketch **Sine Challenge.gsp** to demonstrate this activity for the class.

THE LAW OF SINES (PAGE 35)

Objective: In this activity, students begin by computing sine ratios in two right triangles. They then use Sketchpad to merge the triangles. By studying the resulting triangle, students derive the Law of Sines.

Prerequisites: Students should know how to compute a sine value as the ratio of a right triangle's sides.

Sketchpad Proficiency: Beginner. Students follow step-by-step directions that involve measuring lengths and calculating ratios.

Class Time: 40–50 minutes

Required Sketch: Law of Sines.gsp

SKETCH AND INVESTIGATE

Q1 The sine of $\angle A = CD/AC = CD/b$.

Q2 The sine of $\angle B = FE/BF = FE/a$.

Q3 $CD = b\sin A = a\sin B$

Q4 Rewriting the equality as a proportion yields $\sin A/a = \sin B/b$.

Q5 The sine of $\angle B = AG/BA = AG/c$. The sine of $\angle C = AG/AC = AG/b$.

Q6 $AG = c\sin(B) = b\sin(C)$. Combining this statement with the equality from Q3 yields the complete Law of Sines:

$$\sin\frac{A}{a} = \sin\frac{B}{b} = \sin\frac{C}{c}$$

Q7 $\angle C = 180° - (50° + 70°) = 60°$. Applying the Law of Sines gives $\sin(60°)/1000 = \sin(50°)/BC = \sin(70°)/AC$. Solving for the unknowns yields $BC = 885$ m and $AC = 1085$ m.

THE LAW OF COSINES (PAGE 36)

The standard proof of the Law of Cosines found in most textbooks is a model of economy. A short series of algebraic manipulations yields $c^2 = a^2 + b^2 - 2ab\cos(\theta)$.

In the Pythagorean Theorem, a^2, b^2, and c^2 represent areas of squares. Is there a corresponding geometric interpretation of a^2, b^2, c^2, and the $2ab\cos(\theta)$ term for the Law of Cosines? The standard proof leaves this question unanswered.

This activity presents a visual/algebraic proof of the Law of Cosines that dates from 1931.

Objective: Students build a Sketchpad model that yields the formula for the Law of Cosines. The proof relates c^2, a^2, and b^2 to the areas of three squares and $2ab\cos(\theta)$ to the combined areas of two congruent parallelograms.

Prerequisites: Students should know the definition of cosine and be familiar with the area formula of a parallelogram.

Sketchpad Proficiency: Intermediate. Students follow step-by-step directions for building a model of interconnected squares and parallelograms.

Class Time: 40–50 minutes

Required Sketch: None

Presentation Sketch: Law of Cosines.gsp. Page 3 of the presentation sketch shows a tessellation pattern derived from the Law of Cosines construction.

SKETCH AND INVESTIGATE

Q1 The two parallelograms are always congruent.

Q2 Imagine translating $\triangle DEI$ so that it occupies the space of $\triangle ABC$. Similarly, translate $\triangle ADF$ so that it occupies the space of $\triangle BEH$. All of Region 1 now sits within Region 2, so the two areas are equal. An animated model of these translations appears on page 2 of the presentation sketch.

Q3 The area of square *ACGF* is b^2, the area of square *GHEI* is a^2, and the area of square *ABED* is c^2.

Q4 The area of the parallelogram is bh.

Q5 $c^2 = a^2 + b^2 + 2bh$

Q6 The measure of $\angle BCJ$ is $180° - \theta$.

Q7 $\cos(180 - \theta) = h/a$, so $h = a\cos(180 - \theta)$.

Q8 $h = a\cos(180 - \theta) = -a\cos(\theta)$

Q9 Substituting $h = -a\cos(\theta)$ into the equality from Q5 yields $c^2 = a^2 + b^2 - 2ba\cos(\theta)$.

Q10 The Law of Cosines gives

$$AB^2 = 4^2 + 6^2 - 2(4)(6)\cos(120°)$$

Substituting $\cos(120°) = -0.5$ yields $AB = \sqrt{76}$.

EXPLORE MORE

Q11 Different arguments are possible. Here's one. Based on the fact that $\triangle ABC$ and $\triangle BEH$ are congruent and perpendicular:

$\overline{AC} \cong \overline{CG}$ and $\overline{AC} \perp \overline{CG}$, and $\overline{CG}$ and $\overline{BH}$ are opposite sides of a parallelogram. Therefore $\overline{AC} \cong \overline{BH}$ and $\overline{AC} \perp \overline{BH}$.

Similarly, $\overline{EH} \cong \overline{HG}$ and $\overline{EH} \perp \overline{HG}$, and $\overline{HG}$ and $\overline{BC}$ are opposite sides of a parallelogram. Therefore $\overline{BC} \cong \overline{EH}$ and $\overline{BC} \perp \overline{EH}$.

Because the sides of $\angle ACB$ are perpendicular to the sides of $\angle BHE$, these two angles are congruent, and the two triangles are congruent by SAS.

This implies that $\overline{AB} \cong \overline{BE}$ and also that $\overline{AB} \perp \overline{BE}$, thus establishing that point *E* is one vertex of a square constructed on $\overline{AB}$.

Q12 Quadrilateral *FGID* is a parallelogram. A convincing argument is similar to the preceding argument, in Q11, based on the fact that $\triangle ADF$ and $\triangle DEI$ are congruent and that their corresponding sides are perpendicular.

Q13 When $\angle ACB = 90°$, the two parallelograms disappear. In this case, the sum of the area of the two small squares $(a^2 + b^2)$ is equal to the area of the large square (c^2). Just as before, this can be seen by translating $\triangle DEI$ and $\triangle ADF$ so that they fit inside square *ABED*.

Q14 When point *C* lies along segment *AB*, the picture illustrates $(a + b)^2 = a^2 + b^2 + 2ab$.

UMS OF SINUSOIDAL FUNCTIONS (PAGE 38)

bjective: Students model wave superposition and wave ıckets, analyzing sums of sinusoidal waves.

rerequisites: Students should be familiar with igonometric graphs and identities, as well as period ıd amplitude.

ketchpad Proficiency: Intermediate to Advanced. udents must plot a function, edit a function, create a ırameter, create hide/show and animation buttons, ıd change the properties of an animation button.

ɔu can do this activity at a beginner's level by providing udents with the completed example sketch, **Sinusoidal ım Work.gsp.**

lass Time: 40–50 minutes. You can reduce the onstruction time by providing the students with the ompleted example sketch. If the students do the onstructions, make sure that they save the file with a new ıme when they finish, for use in the Products of nusoidal Functions activity.

equired Sketch: Trig Coords.gsp

xample Sketch: Sinusoidal Sum Work.gsp, Speed of ght.gsp, and **Piano.gsp.**

UMS OF FUNCTIONS

order to simplify this activity, nearly all of the functions ≀ed here are sine functions. You may wish to point out at a cosine graph is merely a horizontal translation of a ıe graph. The function behavior displayed here extends any combination of sine and cosine functions.

Q1

$f(x)$	amp	period	horiz. trans.
$\sin(3x)$	1	$2\pi/3$	0
$4\sin(x)$	4	2π	0
$\sin(x - \pi/2)$	1	2π	$+\pi/2$
$4\sin[3(x - \pi/2)]$	4	$2\pi/3$	$+\pi/2$

Q2 In the function $f(x) = a \cdot \sin[b(x - c)]$, the amplitude is a, the period is $2\pi/b$, and the horizontal translation is c.

Q3

function	period	horiz. trans.
f	2π	0
g	2π	$+\pi/2$
h	2π	$+\pi/4$

The period is the same for all three functions. The horizontal translation of h is the mean of the translations of the component functions. (More data would be needed to establish this as a general result.)

Q4 The length of each wave packet is 2π. The greatest amplitude is very close to 2. (It's not quite 2 because the two waves don't reach their maximum amplitudes at the same x-value.)

Q5

$f(x)$	$g(x)$	$a - b$	packet length
$\sin(6x)$	$\sin(5x)$	1	2π
$\sin(10x)$	$\sin(9x)$	1	2π
$\sin(10x)$	$\sin(9.5x)$	0.5	4π
$\sin(10x)$	$\sin(8x)$	2	π
$\sin(16x)$	$\sin(14x)$	2	π

From the observations on the table, you can infer that the packet length $= \frac{2\pi}{(a - b)}$.

In these five cases, a, b, and $(a - b)$ are all positive. When one or more of these quantities are negative, the packet length calculation must involve absolute values and should therefore be

$$\frac{2\pi}{\left| |a| - |b| \right|}$$

Q6 All three graphs move with the same speed and direction.

Q7 In one function, the changing horizontal translation parameter is t, but in the other, it is $-t$. This causes the horizontal translation to change in opposite directions. The packets are now moving to the right at a faster rate. There is no change in the length of the packets.

Q8 When the component waves move with the same speed and direction, the graph of h simply moves with them. When the components move in opposite directions, h has a much more complicated motion. Although the packets are moving quickly, close inspection shows that individual wave crests are moving very slowly.

Watch one crest rise and fall, alternating between crest and trough. It is created by converging crests and troughs from the component waves. After a certain period of time, 6 f-crests will pass a given point on the x-axis. During that same period, only 5 g-crests will go by. Therefore, the point of convergence must move slowly in order to see the same number of crests pass from each direction.

Q9 Conjectures will vary; the important thing is that students speculate about the result before trying the experiment. Make sure students record their conjectures before actually creating the graph.

Q10 Points on the graph move up and down, but the crests do not move laterally at all. Only the amplitude changes.

PHYSICS CONNECTIONS

Q11 Examples of physical objects that exhibit standing wave behavior are guitar, violin, and piano strings, although such strings vibrate so quickly that they are difficult to observe. You can also demonstrate standing waves with a slinky or with a long pigtail telephone cord, both of which vibrate more slowly.

If a piano is available, playing the lowest C and C# keys at the same time allows students to hear beats clearly. The frequency of low C is 65 hertz, and C# is about 69 hertz, so they produce wave packets with a frequency of about 4 hertz—slow enough to hear each individual packet, or beat, very clearly.

Q12 Although the wave packets may travel faster than light, this effect could be created only by two or more interfering radio signals. The signals themselves are limited to light speed. The sketch **Speed of Light.gsp** demonstrates this effect.

EXPLORE MORE

The example given includes the first three terms of a series that generates a square wave. Each term that is added brings the sum closer to the shape of a square wave. In the limiting case, as the number of terms increases without limit, the plot would become a perfect square wave.

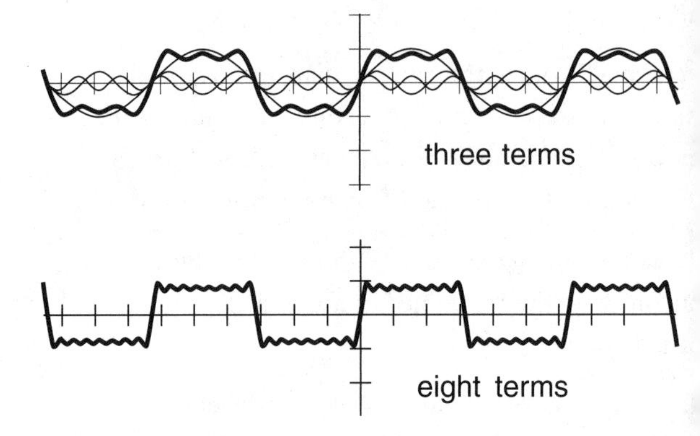

RELATED SKETCH

The sketch titled **Piano.gsp** is a closely related demonstration. It models sound wave superposition. Press the piano keys to see the corresponding sound wave. If two or more notes are pressed, the resulting wave is the sum of the components. Try pressing the lowest C and C# keys at the same time to show a graph of the same beats that can be heard on a real piano.

PRODUCTS OF SINUSOIDAL FUNCTIONS
(PAGE 42)

Objectives: Students graph products of sinusoidal equations and study their properties. This is a continuation of the Sinusoidal Sums activity.

Prerequisites: Students should first complete the Sinusoidal Sums activity. Students should also be familiar with the identities involving $\sin\alpha + \sin\beta$ and $\cos\alpha + \cos\beta$.

Sketchpad Proficiency: Intermediate. You can do this activity at a beginner's level by providing students with the completed example sketch, **Sinusoidal Prod Work.gsp.**

Class Time: 40–50 minutes. You can reduce construction time by having students start with the completed example sketch.

Required Sketch: Students should have available the sketch they created for the previous activity. (If this sketch is not available, students can start from scratch, using the **Trig Coords.gsp** sketch.)

Example Sketch: Sinusoidal Product Work.gsp

INVESTIGATE WAVE PACKETS

Q1 The graph of g has a wave crest or trough corresponding to each crest or trough in h. For one wave packet, they will have the same sign. For the next packet, their signs are opposite. The alternating pattern continues throughout.

Q2 It is a simple, continuous translation. All three graphs move to the right at the same speed. The individual wave crests move in the same way.

Q3 The packets now move to the left, with the same speed and direction as f, which still envelops the packets. The individual crests move to the right, with the same speed and direction as g.

Q4 The amplitude of the envelope is 3, and the period is 2π. The amplitude is the product of the amplitudes of the two components; the period is equal to that of the component with the longer period.

ENVELOPES OF WAVE PACKETS

Q5 The graph of $q(x) = 3\sin(10x)$ does not form an envelope in this case. The coefficient must be associated with the component that has the greatest period.

$h(x)$	envelope $q(x)$
$5\sin(x) \cdot 2\sin(6x)$	$10\sin(x)$
$3\sin(12x) \cdot 3\cos(2x)$	$9\cos(2x)$
$0.5\cos(9x) \cdot 0.8\sin(2x)$	$0.4\sin(2x)$

Q6 The envelope function is $2\cos x$.

SUMMARY

Q7 $h(x) = 2k\sin\left(\frac{a+b}{2}x\right) \cdot \cos\left(\frac{a-b}{2}x\right)$

A function forming the envelope:

$$q(x) = 2k\cos\left(\frac{a-b}{2}x\right)$$

The amplitude of the envelope is $2k$. Its period is $4\pi/(a-b)$. The length of the wave packets is $2\pi/(a-b)$.

EXPLORING THE ROOTS OF QUADRATICS
(PAGE 49)

Objective: Students use linked coordinate systems to explore the relationship between the roots of a quadratic function and the shape of its graph, and to explore the relationship between the roots of a quadratic function and its coefficients. They then connect their results to the quadratic formula.

Prerequisites: Students should be familiar with the relationship between a quadratic's roots and the values of its coefficients. They should also know the quadratic formula.

Sketchpad Proficiency: Beginner. Students use a pre-made sketch.

Class Time: 40–50 minutes

Required Sketch: Quadratic Roots.gsp

This activity is based on an idea of Wallace Feurzeig, Gabriel Katz, Phillip Lewis, and Victor Steinbok.

QUADRATIC ROOTS

Q1 The function $f(x)$ will pass through the origin when one or both of its roots equal 0. This occurs when point R lies on either the r_1- or r_2-axis.

Q2 When the two roots of $f(x)$ are equal, the function will intersect the x-axis at only one point. This occurs when point R lies on the line $r_2 = r_1$.

Q3 The function $f(x)$ is symmetric across the y-axis when its two roots lie on opposite sides of the y-axis, equidistant from the origin. This occurs when point R lies on the line $r_2 = -r_1$.

QUADRATIC COEFFICIENTS

Q4 $r_1 + r_2 = -b$; $r_1 r_2 = c$.

Q5 When point R lies on the r_1-axis, its r_2 value is 0. Thus $b = -r_1$ and $c = 0$. As point R moves along the positive r_1-axis, point P moves along the negative b-axis in the opposite direction and at the same distance from the origin.

Q6 When point R lies on the r_2-axis, its r_1 value is 0. Thus $b = -r_2$ and $c = 0$. So as point R moves along the positive r_2-axis, point P moves along the negative b-axis, at the same distance from the origin.

A CONNECTION TO THE QUADRATIC FORMULA

Q7 When point P lies on the c-axis, its b value equals 0. Thus $r_1 + r_2 = 0$, and $r_1 = -r_2$. The value of c is $r_1 r_2 = -(r_2)^2$, which can never be positive.

Q8 Experimentation suggests that point P lies on the boundary of the traced and untraced regions when point R lies on the line $r_1 = r_2$.

Q9 When $r_1 = r_2$, $b = -2r_1$ and $c = (r_1)^2$.

Q10 Because $b = -2r_1$, $b^2 = 4(r_1)^2 = 4c$. Thus $b^2 = 4c$. This is the equation of the parabola that forms the boundary between the traced and untraced regions of the b-c plane.

Q11 When $b^2 < 4c$, the function $f(x)$ has no real roots. When $b^2 > 4c$, the function $f(x)$ has two distinct real roots.

Q12 The roots of $f(x)$ are

$$\frac{-b \pm \sqrt{b^2 - 4c}}{2}$$

Q13 The function $f(x)$ has two distinct roots when the expression in the square root, namely $b^2 - 4c$, is positive. The roots are identical when $b^2 - 4c = 0$. There are no roots when $b^2 - 4c < 0$.

These results match the answers from Q10 and Q11.

ANALYTIC CONICS (PAGE 51)

Objective: Students build their understanding of analytic representations of conics by making changes to two equation forms and immediately seeing changes in the resulting curve.

Prerequisites: Ideally, students should be familiar with conic sections expressed by equations in standard form and general form. You can use the activity to explore the equation properties before they are introduced, but it may be more effective as a visual reinforcement of concepts that students have already learned.

Sketchpad Proficiency: Beginner. Students need only edit parameters in a completed sketch.

Class Time: 30–40 minutes. The sketch manipulations should not take long, but the questions will take a while to answer. Either part of the activity can stand alone.

Required Sketch: Conics.gsp

CIRCLE, ELLIPSE, AND HYPERBOLA IN STANDARD POSITION

Not all texts use the phrase *standard position* in the same way. In this case, the x- and y-axes are the axes of the conic section. Usually, the denominators are in the form a^2 and b^2. This sketch uses a different (closely related) form in order to make negative terms possible while using only two parameters.

Q1 Ellipse: $A > 0, B > 0$, and $A \neq B$
Circle: $A = B > 0$
Hyperbola: A and B have opposite signs.

Q2 The major axis matches the variable above the larger parameter. For example, if $B > A$, then the y-axis is the major axis, because y is above B. The major radius is the square root of the larger parameter, and the minor radius is the square root of the smaller.

Q3 The transverse axis is determined by the variable above the positive parameter. For example, if $A > 0$, then the x-axis is the transverse axis. The transverse radius is the square root of the positive parameter, and the conjugate radius is the square root of the negative parameter.

Q4 There is no solution if $A = 0$, if $B = 0$, or if A and B are both negative.

Q5 If the axis of a parabola is horizontal or vertical, then either the x^2 term or the y^2 term must have a zero coefficient. That is not possible when the equation is in this form.

GENERAL SECOND-DEGREE EQUATIONS

The general equation used with this file can represent any second-order Cartesian equation in x and y. For simplicity, the xy-coefficient is set to zero for most of the activity, which is how the equation appears in most textbook introductions.

Q6 When $B = 0$, the locus is a conic section and the axes of the conic are either horizontal or vertical.

Q7 The responses in the table can vary endlessly. Below are some correct example responses. As a guide to evaluating the answers, refer to the answers to Q8.

Locus	Equation
Circle	$3x^2 + 3y^2 - 8x + y - 5 = 0$
Ellipse	$2x^2 + 3y^2 - 8x + y - 5 = 0$
Hyperbola	$-2x^2 + 3y^2 - 8x + y - 5 = 0$
Parabola	$3y^2 - 8x + y - 5 = 0$
Line	$-8x + y - 5 = 0$
Intersecting lines	$x^2 - 9y^2 + 4x + 6y + 3 = 0$
Parallel lines	$3y^2 + y - 5 = 0$
Point	$x^2 + 9y^2 + 4x - 6y + 5 = 0$
No solution	$x^2 + 9y^2 + 4x - 6y + 6 = 0$

Encourage students to consider these curves as the intersection of a plane and a cone. How can that intersection create the curve in question? One apparent conflict arises in the case of two parallel lines. Consider a cylinder, the limiting case of a cone when the apex goes to a point at infinity. In this case, a plane can intersect it on two parallel lines.

Q8 Certain other conditions must be satisfied in order to produce real solutions, but these relations hold true:

Circle: $A = C$
Ellipse: $A \neq C$, and A and C have the same sign.
Hyperbola: A and C have opposite signs.
Parabola: $A = 0$ or $C = 0$

Q9 The solution to a second-degree equation is a line when the equation is a quadratic equation of one variable with a double root. Example: $x^2 + 6x + 9 = 0$. Remind students that if a plane is tangent to a cone, the set of intersection points is a line.

Q10 The generalizations do not hold up when $B \neq 0$. In this case, the curve is rotated and interpreting the equation gets much more complicated. This sketch may become useful again later if you pursue this subject.

PARAMETRIC FUNCTIONS (PAGE 53)

Objective: Students explore parametric functions in rectangular and polar coordinates.

Prerequisites: No prior knowledge of parametric functions is necessary. Students should understand the periodic nature of sinusoidal functions.

Sketchpad Proficiency: Beginner. Students only need to edit the existing functions and limit values.

Class Time: 40–50 minutes. For a shorter activity you can leave out the Polar Coordinates section. You can easily extend or abbreviate the Explore More section to fit the curriculum and the class schedule.

Required Sketch: Parametric.gsp

RECTANGULAR COORDINATES

This activity uses θ rather than t as the parameter. This may be confusing to students who have seen only t used as a parameter in the context of parametric functions. This change creates an opportunity to explain that parametric functions only require you to express x and y as functions of some other variable (the parameter), and that it does not matter what you call that other variable.

Q1 When $\theta = 0$, P is at $(2, -1)$.

When $\theta = 1$, P is at $(5, -3)$.

Q2 $x(\theta) = -5 + 7\theta$ and $y(\theta) = 6 - 5\theta$. Other answers are possible.

Q3 $x(\theta) = x_1 + (x_2 - x_1)\theta$

$y(\theta) = y_1 + (y_2 - y_1)\theta$

Q4 The circle has its center at $(0, 0)$ and has a radius of 1. The equation is $x^2 + y^2 = 1$. This equation can be derived from the Pythagorean identity $\sin^2\theta + \cos^2\theta = 1$ by substitution.

Q5 Multiply each parametric function by 3: $x(\theta) = 3\cos\theta$, and $y(\theta) = 3\sin\theta$.

Q6

	Period	Amplitude
x	$\pi/5$	5
y	$2\pi/11$	3

Q7 The function x will complete 10 cycles, while the function y completes 11.

OLAR COORDINATES

ou may need to explain for your students what reflection nd rotation symmetries are.

Q8 6 rotation symmetries, 12 reflection symmetries

Q9 6 rotation, 0 reflection

10 One example:

$r(x) = 4 + \sin 3x$

$\theta(x) = x + (\pi/8)\sin 3x$

XPLORE MORE

ere is a list of the function definitions used for the curves the pictures:

11 $x(\theta) = \theta^2$

$y(\theta) = \theta^3 - \theta$

12 $x(\theta) = 0.5 + \text{round}(\theta) + (\theta - \text{round}(\theta))(\text{sgn}(\theta - \text{round}(\theta)) + 1)$

$y(\theta) = x(\theta + 0.5)$

13 $r(x) = \frac{x}{5} + \sin\left(\frac{\pi}{2} + 2x\right)$

$\theta(x) = \frac{\pi}{12}\sin x$

14 $r(x) = \dfrac{2}{\cos\left(x - \left(\frac{4\pi}{7}\right)\text{trunc}\left(\frac{7x}{4\pi}\right) - \frac{2\pi}{7}\right)}$

$\theta(x) = x$

SURFACES (PAGE 56)

Objective: Students manipulate prepared sketches to view and observe the properties of quadric surfaces, including conic surfaces of revolution and a hyperbolic paraboloid.

Prerequisites: Students should be acquainted with the equations for conic sections.

The sketch traces moving locus objects to generate surface images. This process can be slow, depending on the computer. You can confirm that the computers are fast enough by opening the sketch and pressing the *Paint x Surface* button.

Sketchpad Proficiency: Beginner. Students press buttons and drag controls but perform no constructions or calculations.

Class Time: 30–40 minutes. The activity is divided into two sections that could be presented separately.

Required Sketch: Surfaces.gsp

SURFACES OF REVOLUTION

Q1 The surfaces are tangent. They touch only at the conic section in the x-y plane, described by the equation.

Q2 Answers will vary. For the oblate spheroid, rotate an ellipse on the minor axis. For the prolate spheroid, rotate an ellipse on the major axis. For the sphere, rotate a circle on either axis.

Other suggested questions:

What surface would you generate by rotating on the z-axis?

You cannot generate a cone using this sketch. How could you approximate a cone?

How could you write equations to represent these surfaces?

HYPERBOLIC PARABOLOID

Q3 Cross-sections perpendicular to the x- and y-axes are parabolas. Those perpendicular to the z-axis are hyperbolas. In each case, the equation of the intersecting plane simply sets one variable equal to a constant. Substituting that constant into the hyperbolic paraboloid equation results in a second-degree equation of two variables.

Q4 Giving parameters A and B the same sign will make the horizontal cross-sections ellipses.

Q5 Make $A = B$. You could call it a *circular paraboloid,* but it is normally just called a *paraboloid.* This is a surface of revolution. In fact, it is the conic revolution that was missing from the first part of the activity.

EXPLORE MORE—FUNCTIONS OF TWO VARIABLES

For many students, this will be a first introduction to functions of two variables. This section encourages them to investigate on their own.

The functions graphed in the two images are $f(x, y) = 0.4 \cdot ||x| - |y||$ and $f(x, y) = \sin x + \cos y$. The function that appears when the sketch is first opened is a simulation of ripples propagating from two points.

You cannot generate any of the surfaces of revolution from the first section in their entirety using a function of x and y. This is because the vertical line rule applies, just as it does with a function of one variable.

COMPOUND INTEREST (PAGE 59)

Objective: Students use iterated calculations to compute and plot the value of a compound interest investment. This value is connected to the constant e and the general formula for continuous compounding.

Prerequisites: Students do not need to understand any of the derivations of e, but they should be familiar with exponential functions in the form ae^{bx}.

Sketchpad Proficiency: Intermediate. The activity is heavy on calculations and plotting. Students must be able to create and edit parameters and calculations and to plot coordinates and functions. There is one iteration.

You could give the activity to beginners by providing them with the completed sketch, but students may lose some understanding if they do not go through the calculations themselves. You could also use the finished sketch as a presentation.

Class Time: 30–40 minutes. Expect a lot of disparity in their rate of progress. Experienced Sketchpad users with good keyboard skills could finish in less than 15 minutes.

Example Sketch: Compound Interest Work.gsp

A SIMPLISTIC INVESTMENT

2. Students should calculate $t + 1/k$ at this step. This result is revealed several steps later.

3. Students should calculate $P \cdot (1 + 1/k)$. This factored form is revealed several steps later.

Q1 The points represent time and current value of the investment. The first point is at inception ($t = 0$). The second point is at the end of the first compounding period.

5. Hiding the points results in a thinner, more precise graph, showing only the iterated segments.

Q2 For quarterly compounding, k must be 4.

Quarterly, \$2.44; monthly, \$2.61; weekly, \$2.69; daily, \$2.71.

Q3 The limit is approximately 2.71828—the value e. Daily compounding reaches 2.71457, and hourly compounding reaches 2.71813, so students are not likely to approximate the value more closely than the nearest hundredth, or perhaps thousandth. Compounding with 100,000 periods (about every five minutes) results in a value of 2.71827.

A MORE REALISTIC INVESTMENT

10. The factored form is $P(1 + r/k)$.
11. The *depth* calculation should be $term \cdot depth - 1$.

Q4 Annually, $150.37; daily, $152.95.

Q5 In the equation $P(1 + r/k)^{kx} = ae^{bx}$, a must equal P. You may write an exponential function in any number of forms, using any positive real number other than 1 as the base, but the coefficient is always the same.

Q6 $A(x) = 100e^{0.085x}$

Q7 $A(x) = Pe^{rx}$

Another representation of the concept that was just observed:

$$\lim_{x\to\infty}\left(1 + \frac{r}{x}\right)^x = e^r$$

SLOPES OF EXPONENTIAL FUNCTIONS (PAGE 62)

Objective: By observing graphs, students make inferences about the relationship between exponential functions and their slopes.

Prerequisites: Students must understand basic properties of exponential functions in the form $f(x) = a^x$. There are references to the derivative of the function, but students need only understand that the derivative describes the slope of a function. It is not necessary for them to know the limit definition.

Sketchpad Proficiency: Intermediate. There are only a few Sketchpad tasks to perform. Most of the activity involves manipulating a graph and making observations.

Class Time: 30 minutes

Required Sketch: Exponential Slope.gsp

Example Sketch: Exponential Slope Work.gsp

SKETCH AND INVESTIGATE

Sketchpad is capable of computing the derivatives of many functions, including the exponential functions used here. However, the **Derivative** tool is not mentioned, because it would undermine the point of this particular activity.

Q1 The function $f(x) = a^x$ is undefined for $a < 0$.

For $a = 0$, it is defined only on the domain $x > 0$, in which case it is a constant function with slope 0.

For $0 < a < 1$, the slope is negative everywhere.

For $a = 1$, it is a constant function with slope 0.

For $a > 1$, the slope is positive everywhere.

Q2 The graphs coincide for $a = e$. This shows that the function $f(x) = e^x$ is its own derivative. If students do not recognize the connection with e, they should still be able to say the same thing about the approximation $f(x) = 2.72^x$.

Q3 Students should observe that

$$\ln a = \frac{Slope(j)}{f(x_A)}$$

and substitute that into the equation.

$$\ln a = \frac{Slope(j)}{f(x_A)} = \frac{f'(x_A)}{f(x_A)} = \frac{f'(x)}{f(x)}$$

$$\frac{f'(x)}{f(x)} = \ln a$$

$$f'(x) = \ln a f(x)$$

Thus, the derivative of a^x is $\ln(a)a^x$.

EXPLORE MORE

Techniques from the activities Translation of Functions and Dilation of Functions will be helpful here. Functions that are equal to their own derivatives include vertical stretches (be^x) and horizontal translations (e^{x-c}) of e^x. Any combination of these transformations will work. The general solution is be^{x-c}.

The constant function with this property is simply the special case of the above solution where $b = 0$. This is equivalent to the constant function $f(x) = 0$.

A SEQUENCE APPROACH TO LOGS (PAGE 64)

Objective: Students graph geometric sequences against arithmetic sequences to obtain good approximations of log curves.

Prerequisites: Students should be familiar with logarithmic functions and the identity $\log_c r^n = n \log_c r$.

Sketchpad Proficiency: Beginner. Students follow step-by-step iteration directions. They should be comfortable using Sketchpad's Calculator.

Class Time: 40–50 minutes

Required Sketch: Logs.gsp

Example Sketch: Logs Work.gsp

This activity is based on an idea of David Dennis.

EXAMINE A TABLE

Q1 When $x = r^2$, $\log_c x = \log_c r^2 = 2\log_c r = 2d$. This same reasoning applies more generally for $x = r^n$: $\log_c x = \log_c r^n = n\log_c r = nd$.

MANIPULATE THE CURVE

Q2 If $y = \log_c x$ passes through (4, 2), then $2 = \log_c 4$. Thus $c^2 = 4$ and $c = 2$.

11. To graph $y = \log_2 x$, rewrite the equation as $y = \log x / \log 2$.

EXPLORE MORE

If $2 = \log_c 9$, then $c = 3$. Thus $y = \log_3 x = \log x / \log 3$.

SEMILOG GRAPHS (PAGE 66)

Objective: Students explore a logarithmic scale and a semilog grid, and use the grid to investigate exponential functions and to solve a problem involving exponential decay.

Prerequisites: Students should first have a good understanding of logarithm properties and exponential equations.

Sketchpad Proficiency: Intermediate. There is extensive use of custom tools, but otherwise the activity is basic.

Class Time: 50–60 minutes. The activity is divided into four sections. If you want to do them in two separate class periods, the best break would be between the second and third sections.

Required Sketch: Semilog.gsp

Presentation Sketch: Semilog Work.gsp

THE LOGARITHMIC SCALE

Q1 The logarithm of negative numbers or zero is undefined. Alternatively, students can point out that even though numbers get smaller and smaller as you go down the vertical scale, they never reach zero.

Q2 The value of $y_{A'}$ is y_A raised to the power of the scale factor. This follows from the logarithm property that

$$\log a^b = b \log a$$

Q3 The value of y_D is the product of y_B and y_C. This follows from the property that

$$\log(ab) = \log a + \log b$$

SEMILOG GRAPHS

Q4 The y-coordinate is consistent with the grid and is positive everywhere. Expect some students to say that it is zero near the bottom. This is only because of rounding. To see a more precise (nonzero) value, change the properties of the y-coordinate measurement by setting the value precision to hundred thousandths.

Q5 The point will travel along a logarithmic curve. It will be easier to see if the point is traced. Try starting the animation from different locations.

EXPONENTIAL EQUATIONS

Q6 The graph is a line. Changing parameter a translates it vertically. Changing parameter b changes the slope, but does not change the y-intercept.

Q7 The graphs of exponential equations of the form $y = ab^x$ are straight lines. Students should be made aware that this is not true of all exponential equations, for example, equations of the form $y = ab^x + c$.

EXPONENTIAL DECAY

Q8 The coordinates of point P are (12.32, 15.00), meaning that after 12.32 years, there will be 15 g of tritium remaining. The half-life of tritium is 12.32 years.

PRESENT

The three forms of the exponential function available in the custom tools are $y = ae^{bx}$, $y = a \cdot 10^{bx}$, and $y = ab^x$.

There is also a tool that will return the exponential function. You can copy this function to a separate page and use it to plot the graph on a linear grid.

RELATED SKETCHES

Semilog Complete.gsp

This is the same file as **Semilog.gsp,** but with some more advanced features displayed. It could be useful for other activities or for pasting semilog graphs into worksheets, tests, and so forth.

LOG-LOG GRAPHS (PAGE 70)

Objective: Students explore graphs of power functions on a log-log grid, and use the grid to solve an interesting statistical problem.

Prerequisites: Students must understand linear graphing and basic logarithm properties.

Sketchpad Proficiency: Intermediate. Custom tools are used. An extension directs students to construct a graph on a linear grid with little guidance.

Class Time: 40–50 minutes. You could leave the presentation extension for another day or eliminate it.

Required Sketch: Log-Log.gsp

Presentation Sketch: Log-Log Work.gsp

SKETCH AND INVESTIGATE

Q1 This particular graph appears just as it would on a linear scale. It is a line through point (1, 1) with slope 1.

Q2 The graph of any function in the form $f(x) = ax^b$ is a line. Parameter a must be greater than zero, but b can be any real number.

Q3 y-intercept $= a$; slope $= b$

Q4 $\log(y/a) = \log x^b$

$y/a = x^b$

$y = ax^b$

This shows that any non-vertical line on the log-log grid represents a power function.

APPLICATION

The application seeks to approximate the relationship between mass and strength so that a more honest comparison can be made between the strength of an ant and that of a human.

The subject can be the topic of interesting discussion. Contrary to sci-fi movie wisdom, a human-sized ant would be hard put to lift its own head. (In fairness, if we were shrunk down to its size, our own anatomy would present great problems too.)

Q5 The slope would be 1, because the function would be a direct variation, $f(x) = ax^1$. Clearly, this is not the case.

Q6 Answers will vary. Some typical responses:

Men: $15.73x^{0.59}$

Women: $10.69x^{0.62}$

Q7 Use the ratio $f(0.0001)/0.0001$, because you must convert the units to kilograms. Men would lift about 500 to 600 times their weight, and women would lift about 300 to 400 times theirs. The precise answers will vary. Minor changes in the best-fit lines have a great effect on the answers. In any case, it would appear that human strength would surpass the ants' by far.

PRESENT

When copying objects to the other page, students may transfer certain parent objects as well. They should hide these, but not delete them. Students will have an easier time plotting the functions if they first scale the axes to an appropriate size. A rectangular grid works well here.

You may ask students to consider why the mass/strength relationship is not a direct variation. The athlete's mass is proportional to the volume of his or her body (3D), whereas the strength of a muscle is roughly proportional to its cross-sectional area (2D). Analysis might predict that strength is proportional to mass to the power 2/3.

This model is very simplistic. There are other factors to consider. Besides the barbell, the athletes have the burden of their own weight. Also, it is not realistic to assume that athletes of different sizes have the same proportions.

RELATED SKETCHES

Log-Log Complete.gsp

This is the same file as **Log-Log.gsp,** but with some more advanced features displayed. You could use it for other activities, or for pasting log-log graphs into worksheets, tests, and so forth.

HE LOGISTIC FUNCTION (PAGE 73)

bjective: Students build a Sketchpad model of a opulation for which the growth is restrained by some ctor. In the process, students explore the sensitivity of the ng-term behavior to the initial size of the population and the parameters that determine the population growth.

rerequisites: Students should be familiar with athematical models of exponential growth and decay.

ketchpad Proficiency: Intermediate. Students must know ow to plot functions and how to do simple geometric onstructions such as segments and parallel lines. The ctivity uses the **Iterate** command, giving specific irections for doing so.

lass Time: 40–50 minutes

equired Sketch: None

xample Sketch: Logistic Work.gsp

EVELOP THE EQUATION

Q1 The size of the next generation will be kp.

Q2 When $n = p$, the limiting factor is 0, so the size of the next generation will be 0. The entire population will die.

Q3 When p is very small relative to n, the limiting factor is very close to 1, so it has no effect on $f(p)$. The value of $f(p)$ is determined entirely by the growth factor.

Q4

p	$f(p)$
5,000	7,462.5
50,000	71,250
500,000	375,000

Q5 You can write the limiting factor as $(1 - x)$. $g(x) = kx(1 - x)$

IRST GENERATION

Q6 The value of x_Q is equal to the value of the function at x_P because the parallel line is constructed through a point whose y-value is $f(x_P)$, so the equation of the line is $y = g(f(x_P))$. At point Q, where this parallel intersects the line $y = x$, the x- and y-values of Q must be equal, so $x_Q = y_Q = g(x_P)$.

NEXT GENERATIONS

Q7 When P is at 0.5 and $k = 1$, the population gets smaller and smaller, eventually approaching zero.

Q8 The initial position of P makes no difference, because when $k = 1$, the population can never grow but can only shrink slightly from one generation to the next. Such a population must eventually die out.

Q9 When you change the value of k to 2.5, the population stabilizes at 0.60. This long-term result does not depend on the initial position of point P.

Q10 When $k = 3.1$, the population alternates between 0.56 and 0.76 in alternate generations. When the population is 0.56, the growth factor predominates and the size of the next generation increases to 0.76. When the population is 0.76, the limiting factor predominates and reduces the next generation to a size of 0.56. The long-term result does not depend on the initial value of P.

Q11 Answers will vary. The two-generation cycle described in Q10 begins at about $k = 3.0$ and begins to break down around $k = 3.45$, becoming a four-generation cycle. By $k = 3.55$, the situation has further deteriorated to an eight-generation cycle. As k continues to increase, the pattern becomes increasingly chaotic and is very sensitive to small changes in the value of k.

To summarize, long-term population size is very stable for values of k from 0 to 3. Values slightly above 3 result in bi-stable long-term behavior, in which the population alternates between two specific sizes. As the value of k gets closer to 4, even this bi-stable behavior breaks down, and the long-term behavior becomes more and more complex, and more and more sensitive to small changes in k. Chaos theory is the study of such systems, in which small changes in parameters produce large changes in behavior.

EXPLORE MORE

The continuous form of the logistic function gives the same long-term behavior as the logistic map when k is less than 3. For instance, when $k = 2$, the long-term population is 0.5, using either the discrete or continuous form. The value of c is related to the initial population.

Page 5 of **Logistic Work.gsp** uses a related equation that depends explicitly on the initial population.

The continuous form does not show the instability of the discrete form. This is because the instability is caused by the discrete jumps from one generation to the next, jumps that do not occur with a continuous function. In this way the discrete formulation may be a better model for many natural phenomena.

LINEAR REGRESSION (PAGE 79)

Objective: Students manipulate scatter plots and experiment with best-fit lines and acquire an intuitive feel for regression lines, residuals, and least squares analysis.

Prerequisites: Students need little understanding of statistics for this activity. They should be familiar with the idea of fitting a line to a scatter plot.

Sketchpad Proficiency: Intermediate. The sketch is nearly complete, but students will have to perform some calculations.

Class Time: 30–40 minutes. Students can complete the steps of the activity in only a few minutes, but should have time to experiment. Also, they should have time to discuss some open-ended questions.

Required Sketch: Linear Regression.gsp

Example Sketch: Linear Regression Work.gsp

SKETCH AND INVESTIGATE

Q1 The segment lengths correspond to the magnitude of the residuals.

Q2 The blue segments are positive, and the red negative. Students should understand that the residual has a sign.

Q3 Precise answers will vary. Using this method, it should be possible to get a total residual of nearly zero for any scatter plot. Simply draw any non-vertical line. Translate it up or down to adjust the total.

Q4 You can adjust the line so that it generates a small sum but fits the data badly. Even when the total is zero, individual residuals may be large, so this is clearly not a promising solution.

Q5 It will help to square the residuals, because the squares will all be positive.

Q6 The area of each square is equal to the square of its corresponding residual.

Q7 Answers will vary. Students are likely to predict that the best-fit line will be nearly vertical.

Q8 The best-fit regression line is actually likely to be closer to horizontal than vertical. This question helps students realize that the magnitude of the residual is not the distance from a point to the line. It is the *vertical* distance.

In fact, a vertical scatter indicates no close data correlation at all, so fitting a line to it might not be realistic.

Q9 Even after deriving the least squares regression line, students can probably find an alternate line that will do a better job of minimizing the sum of the absolute values. These two lines are usually very close.

The rationale for using squares lies mainly with the cumbersome nature of the absolute value function. It has a *piecewise* definition—it has one definition when $x \geq 0$ and a different definition when $x < 0$. Such functions do not work well with formulas, especially with large data sets.

The *Seek Least* $|r|$ button does a fair job of minimizing the sum of absolute values, but it is executing an algorithm, not a formula, and does not give an exact result. Points Q and R constantly adjust their motions as thousands of calculations are being performed, and this is for a data set of only eight elements.

Remind students that the foundations of statistical science were laid during the not-so-distant era of slide rules and logarithm tables.

ENTER DATA

Q10 $y = 7.56x + 21.33$

Answers may vary somewhat.

EXPLORE MORE

Encourage students to adjust the line in Sketchpad by eye before checking it on a calculator. They may be surprised to find how close they can come.

WAIT FOR A DATE (PAGE 81)

Objective: Students use a pre-made Sketchpad model to gather sample data for a probability problem. By viewing the data as points in a plane, students uncover a geometric pattern that allows them to compute a precise probability.

Prerequisites: Students should be familiar with basic notions of probability.

Sketchpad Proficiency: Beginner. Students work with a pre-made sketch.

Class Time: 40–50 minutes

Required Sketch: Wait for a Date.gsp

MODEL IN ONE DIMENSION

Q1 If you and your friend wait more than 10 minutes for each other, it is more likely you will meet.

MODEL IN TWO DIMENSIONS

Q2 Points on the diagonal segment represent instances in which you and your friend arrive at the exact same moment. Any placement of point P on or very close to this segment guarantees you'll meet. Locations 2 and 3 both satisfy this condition.

Q3 The green points cluster entirely along a diagonal strip, whereas the red points cluster entirely in two congruent triangular regions on either side of the strip. In this illustration, G = green and R = red.

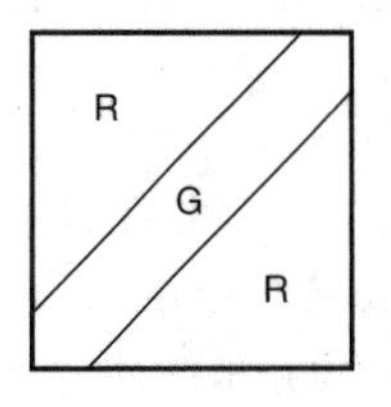

Q4 The diagonal segment connecting (12:00, 12:00) and (1:00, 1:00) represents the set of points for which you and your friend arrive at the exact same time. This is the same diagonal segment that appears in Q2.

Q5 Pick any point on the diagonal segment from Q4, for instance (12:30, 12:30). If your friend arrives 10 minutes after you do, this point is translated vertically by 10 minutes to (12:30, 12:40).

The same pattern holds for every point on the diagonal segment. The collection of all such points is a segment that is a vertical translation by 10 of the diagonal segment. The exact coordinates of the line run from (12:00, 12:10) to (12:50, 1:00). This new segment represents one border of the green region shown in Q3.

Q6 The same reasoning from Q5 applies here. Translating the diagonal segment from Q4 by −10 minutes vertically represents those times when your friend arrives 10 minutes before you do. The exact coordinates of this line run from (12:10, 12:00) to (1:00, 12:50). This segment is a border of the green region shown in Q3.

Q7 If the length of the horizontal and vertical axes is 60 (representing 60 minutes), then the base and height of each red right triangle is 50. The area of both triangles combined is $50 \times 50 = 2500$. This is 69% of the square, leaving 31% for the green points.

Q8 The portion of the square filled with green points represents the probability that you and your friend will meet. Thus there is a 31% chance of your meeting.

EXPLORE MORE

Q9 Answers will vary.

Q10 The probability that you and your friend will meet is $1 - (1 - t/60)^2$.

Q11 The lower triangle has base and height of 55 and an area of 1512.5. The upper triangle has base and height of 45 and an area of 1012.5. The red area (the sum of the triangle areas) is 2525, which is 70% of the square. Therefore the probability that you and your friend will meet is 30%. In general, if one person is willing to wait t_A minutes and the other is willing to wait t_B minutes, the probability is

$$1 - \frac{1}{2}\left(1 - \frac{t_A}{60}\right)^2 - \frac{1}{2}\left(1 - \frac{t_B}{60}\right)^2$$

FITTING POLYNOMIAL FUNCTIONS (PAGE 83)

Objective: Students use matrix algebra to fit a polynomial function to three arbitrary points on the *x*-*y* plane.

Prerequisites: Students will need to be familiar with the concepts of matrix multiplication and the inverse matrix. Custom tools perform the calculations, so it is not necessary for the student to actually process all of the numbers.

Sketchpad Proficiency: Intermediate

Class Time: 30 minutes. The suggested extensions, four points or five points, are no more complex conceptually, but require the manipulation of many more numbers.

Required Sketch: Polynomial Fit.gsp

Example Sketch: Polynomial Fit Work.gsp

Q1 Having two points with the same *x*-coordinate would conflict with the vertical line rule. No function of any kind can map one value of *x* to two different values of *y*.

SKETCH AND INVESTIGATE

2. The equivalence of the two matrix equations shown may not be obvious to students. The margin note suggests that they think of these equations as though they were normal algebraic equations:

$$M \cdot a = y \text{ is equivalent to } M^{-1} \cdot y = a$$

Q2 Consider a case of $n - 1$ points fitting an $n - 2$ degree polynomial function. If an *n*th point were introduced and happened to fall on the same curve, then the same $n - 2$ degree function would fit the *n* points. In the case of three points, this happens when they are collinear.

Q3 There are infinitely many. You can demonstrate this in the completed sketch by dragging point 3 while the other two points remain fixed. The curve changes continuously but always fits the two fixed points. Thus, infinitely many second-degree polynomials fit two points.

Q4 There is no inverse matrix in that case, because the calculations would result in division by zero. In practice, because of processing limitations, this may only cause some of the inverse matrix elements to have very large numbers.

PRESENT

This section suggests using a custom template for the presentation. Though this is an advanced feature, it is not hard to use. Go to the Help topic listed for more details.

EXPLORE MORE

The extension suggests that students create separate pages for cases with four points or five points. As an alternative, they could use the existing points as the first three points for the four-point case, and so on.

When fitting a polynomial to more than three points, students must be careful to organize the workspace effectively. The five-point case could turn the screen into a sea of numbers.

The sketch has quite a number of tools not used in the activity, including determinant and adjoint matrix tools. More advanced linear algebra students could use these tools to compute the inverse matrix or apply Cramer's Rule.

RELATED SKETCHES

The file **Polynomial Fit Work.gsp** has the completed activity for cases $n = 1, 2, 3, 4$, and 5.

MATRIX TRANSFORMATIONS (PAGE 87)

Objective: Students perform elementary coordinate transformations using matrices and vector addition.

Prerequisites: Students should be familiar with matrix multiplication.

Sketchpad Proficiency: Intermediate. Most of the tasks involve calculations and the use of a custom tool.

Class Time: 30–40 minutes

Required Sketch: Matrix Transformations.gsp

Example Sketch: Matrix Transformations Work.gsp

TRANSFORM A POLYGON

Q1 This first matrix creates a 180° rotation about the origin point O. It might also be called a dilation by scale factor -1 or a point reflection.

Q2 **A:** reflection across the y-axis

B: reflection across the x-axis

C: reflection across the line $y = x$

D: dilation about O by scale factor r

E: horizontal stretch from y-axis by scale factor r

F: vertical stretch from x-axis by scale factor r

G: rotation clockwise by angle θ

H: rotation counterclockwise by angle θ

COMBINE TRANSFORMATIONS

Q3 To create a dilation about point Q, students do not need to start over. They can just change the matrix to the dilation matrix:

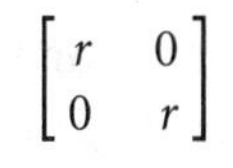

$$\begin{bmatrix} r & 0 \\ 0 & r \end{bmatrix}$$

EXPLORE MORE

Students will need to know how to multiply matrices. The first matrix product, **GE,** is a rotation followed by a stretch in the x direction. The product **EG** is the stretch followed by the rotation.

The results are different, because matrix multiplication is not commutative. In the first case, the image changes shape as angle θ changes. In the second case, it remains rigid as θ changes.

VECTOR OPERATIONS (PAGE 89)

Objective: Students observe geometric interpretations of a dot product and a vector projection, and then compute and plot a vector projection.

Prerequisites: Students should know how to compute a dot product from vector components and from vector lengths and intersection angle. They will be guided through the vector projection calculations, but they may benefit by knowing what a vector projection is before beginning.

Sketchpad Proficiency: Intermediate. Students should be able to use the Calculator and prepared custom tools with little guidance.

Class Time: 30–40 minutes

Required Sketch: Vector Operations.gsp

Example Sketch: Vector Operations Work.gsp

GEOMETRY

Q1 $a_1 = x_R - x_Q$
$a_2 = y_R - y_Q$

Q2 Translating a vector has no effect on its components.

2. The formula is $\mathbf{a} \cdot \mathbf{b} = a_1b_1 + a_2b_2$. If students don't already know this formula, they should just use the **Dot Product** tool. They will see the formula in step 14.

Q3 The translation has no influence on the dot product.

Q4 The absolute value of the dot product increases when you make one of the vectors longer and decreases when you make one of the vectors shorter.

Q5 The dot product is zero when the vectors are perpendicular. It has a maximum value when the vectors have the same direction and a minimum value when the directions are opposite.

Q6

	Translate	Change Direction	Change Magnitude
a	*N*	*Y*	*Y*
b	*N*	*Y*	*N*

CALCULATIONS

Q7 Point *A*^ is on ray *OA*. Its distance from the origin is 1.

Q8 When $\angle AOB$ is obtuse, **p** is in the opposite direction from **â.**

Q9 Look again at the formula that was given for $|\mathbf{p}|$. It actually works only when $\angle AOB$ is acute. In the case of an obtuse angle, the formula gives a negative value for $|\mathbf{p}|$. That is also the only case in which vectors **â** and **p** have opposite directions. Thus the vector is effectively turned around.

Q10 The three calculations of the dot product always agree. In fact, the **Dot Product** tool uses the same formula as step 14: $\mathbf{a} \cdot \mathbf{b} = a_1b_1 + a_2b_2$.

ATRIX PRODUCTS (PAGE 92)

bjective: Students create complex transformations using ne product of several transformation matrices.

rerequisites: Students should be familiar with 2 × 2 atrices for rotation, dilation, and reflection.

ketchpad Proficiency: Intermediate. Most of the work volves editing parameters and calculations.

lass Time: 50–60 minutes. The activity is divided into ree sections, which may be performed together or on parate days.

equired Sketch: Matrix Product.gsp

xample Sketch: Matrix Product Work.gsp

hroughout the activity students are instructed to keep a ritten record of the matrices they edit. The written record ill help them organize their thoughts and will make it easier r you to evaluate the work and help students locate errors.

RANSLATION

he matrix in response to Q1 is an important first step. tudents must understand how to convert a parameter into calculation and use the coordinate measurements from ne sketch.

Q1 This is the matrix definition:

$$\begin{bmatrix} 1 & 0 & x_R - x_Q \\ 0 & 1 & y_R - y_Q \\ 0 & 0 & 1 \end{bmatrix}$$

OTATION/DILATION

Q2 The matrix product **CBA:**

$$\begin{bmatrix} 1 & 0 & x_Q \\ 0 & 1 & y_Q \\ 0 & 0 & 1 \end{bmatrix}\begin{bmatrix} \cos\theta & -\sin\theta & 0 \\ \sin\theta & \cos\theta & 0 \\ 0 & 0 & 1 \end{bmatrix}\begin{bmatrix} 1 & 0 & -x_Q \\ 0 & 1 & -y_Q \\ 0 & 0 & 1 \end{bmatrix}$$

Q3 Here is **CBA.** Only matrix **B** was changed.

$$\begin{bmatrix} 1 & 0 & x_Q \\ 0 & 1 & y_Q \\ 0 & 0 & 1 \end{bmatrix}\begin{bmatrix} r & 0 & 0 \\ 0 & r & 0 \\ 0 & 0 & 1 \end{bmatrix}\begin{bmatrix} 1 & 0 & -x_Q \\ 0 & 1 & -y_Q \\ 0 & 0 & 1 \end{bmatrix}$$

REFLECTION

The reflection line itself is transformed to make it easier to follow each step of the transformation. The line is translated and then rotated so that it coincides with the x-axis. The image of point P is then reflected on the x-axis. Finally, the first two transformations are reversed so that the line is sent to its original position.

Sine and cosine calculations are used in step 12 to demonstrate how the transformation matrix can be calculated given only the coordinates of the points involved.

Q4 Matrix product **EDCBA:**

$$\begin{bmatrix} 1 & 0 & x_Q \\ 0 & 1 & y_Q \\ 0 & 0 & 1 \end{bmatrix}\begin{bmatrix} \cos\theta & -\sin\theta & 0 \\ \sin\theta & \cos\theta & 0 \\ 0 & 0 & 1 \end{bmatrix}\begin{bmatrix} 1 & 0 & 0 \\ 0 & -1 & 0 \\ 0 & 0 & 1 \end{bmatrix}$$

$$\begin{bmatrix} \cos\theta & \sin\theta & 0 \\ -\sin\theta & \cos\theta & 0 \\ 0 & 0 & 1 \end{bmatrix}\begin{bmatrix} 1 & 0 & -x_Q \\ 0 & 1 & -y_Q \\ 0 & 0 & 1 \end{bmatrix}$$

EXPLORE MORE

You can see the rolling circle suggested in this section on the Cycloid page of the example file. It requires considerable independent thought, but the transformation uses only two matrices.

Point B is a free point, and O is the origin. Measurement OB is the coordinate distance. Rotate by angle $-r$ (radians). Translate by vector $(OB \cdot r, OB)$. Transform points O and B, and construct a circle centered on O', through B'.

$$\mathbf{BA} = \begin{bmatrix} 1 & 0 & OB \cdot r \\ 0 & 1 & OB \\ 0 & 0 & 1 \end{bmatrix}\begin{bmatrix} \cos r & \sin r & 0 \\ -\sin r & \cos r & 0 \\ 0 & 0 & 1 \end{bmatrix}$$

COORDINATES IN THREE DIMENSIONS (PAGE 95)

Objective: Students explore coordinates in three dimensions using rectangular, cylindrical, and spherical coordinate systems.

Prerequisites: This activity assumes an understanding of rectangular and polar coordinates in two dimensions. Students should also have a good foundation in trigonometry.

Sketchpad Proficiency: Intermediate. Students edit parameters and use custom tools supplied with the sketch. In an extension, they create parameters, perform calculations, and use the results in the tools.

Class Time: 30–40 minutes for the main part of the activity. The extension at the end could take much longer. It might be advisable to leave the calculations as a take-home assignment and do the plotting another day.

Required Sketch: Coordinates 3D.gsp

Example Sketch: Coordinates 3D Work.gsp

The custom tools used in this activity accept dimensionless parameters or measurements. They are interpreted as radians and centimeters, based on the file's Preferences settings. You can change these defaults by changing the Preferences. Inches, pixels, and directed degrees are fine, but do not use degrees.

RECTANGULAR COORDINATES

Q1 When one of the coordinates is changed, the point moves in a straight line parallel to the corresponding axis.

CYLINDRICAL COORDINATES

Q2 The motion of the point depends on which coordinate is being changed.

r: in a line away from or toward the z-axis

θ: in a circular path around the z-axis

z: in a line parallel to the z-axis

Q3 Any number of (x, y, z) points can appear to match the (r, θ, z) point if they lie on the line joining the plotted point and the projection point. Changing the perspective will show that these points do not match.

Q4 The correct answer is $(3\cos(1.4), 3\sin(1.4), -2.45)$ or about $(0.51, 2.96, -2.45)$.

SPHERICAL COORDINATES

Q5 ρ: in a line away from or toward the origin

θ: in a circular path around the z-axis

ϕ: in a circular path around the origin

Q6 $(\rho\cos\theta\sin\phi, \rho\sin\theta\sin\phi, \rho\cos\phi)$ is about $(1.30, -2.19, -2.78)$.

EXPLORE MORE

Completing all three versions of the cube would require computing and plotting 24 points and connecting them with 36 line segments. To avoid the tedium, consider assigning only one version of the cube. If more are plotted they should be done on separate pages or separate files.

Answers may vary, depending on the position and orientation chosen for the cube. In these solutions, the cube is centered on the origin and each axis runs through the center of two faces.

In all three cases, there is quite a lot of repetition of numbers. When using the custom tools to plot the points, it is possible to use the same measurement or parameter for several points, or even to use it more than once on the same point. For example, the first rectangular coordinate point would be plotted by clicking the same measurement three times.

x	y	z
$s/2$	$s/2$	$s/2$
$-s/2$	$s/2$	$s/2$
$-s/2$	$-s/2$	$s/2$
$s/2$	$-s/2$	$s/2$
$s/2$	$s/2$	$-s/2$
$-s/2$	$s/2$	$-s/2$
$-s/2$	$-s/2$	$-s/2$
$s/2$	$-s/2$	$-s/2$

r	θ	z
$(\sqrt{2}/2)s$	$\pi/4$	$s/2$
$(\sqrt{2}/2)s$	$3\pi/4$	$s/2$
$(\sqrt{2}/2)s$	$5\pi/4$	$s/2$
$(\sqrt{2}/2)s$	$7\pi/4$	$s/2$
$(\sqrt{2}/2)s$	$\pi/4$	$-s/2$
$(\sqrt{2}/2)s$	$3\pi/4$	$-s/2$
$(\sqrt{2}/2)s$	$5\pi/4$	$-s/2$
$(\sqrt{2}/2)s$	$7\pi/4$	$-s/2$

ρ	θ	ϕ
$(\sqrt{3}/2)s$	$\pi/4$	$\cos^{-1}(\sqrt{3}/3)$
$(\sqrt{3}/2)s$	$3\pi/4$	$\cos^{-1}(\sqrt{3}/3)$
$(\sqrt{3}/2)s$	$5\pi/4$	$\cos^{-1}(\sqrt{3}/3)$
$(\sqrt{3}/2)s$	$7\pi/4$	$\cos^{-1}(\sqrt{3}/3)$
$(\sqrt{3}/2)s$	$\pi/4$	$\pi - \cos^{-1}(\sqrt{3}/3)$
$(\sqrt{3}/2)s$	$3\pi/4$	$\pi - \cos^{-1}(\sqrt{3}/3)$
$(\sqrt{3}/2)s$	$5\pi/4$	$\pi - \cos^{-1}(\sqrt{3}/3)$
$(\sqrt{3}/2)s$	$7\pi/4$	$\pi - \cos^{-1}(\sqrt{3}/3)$

PARAMETRIC FUNCTIONS IN THREE DIMENSIONS (PAGE 98)

Objective: Students use parametric functions to draw curves in rectangular (x, y, z) coordinates. An open-ended extension uses cylindrical and spherical coordinates.

Prerequisites: Students should be familiar with simple parametric functions in the plane, perhaps by having completed the Parametric Functions activity in Chapter 4. If students pursue the cylindrical/spherical coordinate extension, they should first complete the Coordinates in Three Dimensions activity in this chapter.

Sketchpad Proficiency: Beginner. Students edit parameters and function definitions, but perform no constructions.

Class Time: 30–50 minutes. The questions require only short answers, but the cylindrical and spherical challenges are open-ended and can take some time.

Required Sketch: Parametric 3D.gsp

Example Sketch: Parametric 3D Work.gsp

PARAMETRIC LINES

If the class has studied vectors, point out to them that this is similar to the vector definition of a line.

Q1 The functions describe a line through the origin. The range limits make it a line segment with the origin as one endpoint.

Q2 When $\theta = 0$, the blue point is at P_1. When $\theta = 1$, it is at P_2. The parametric functions describe a line through these two points.

Q3 Set the lower and upper limits to 0 and 1.

THE HELIX

Q4 Students' answers will vary. It's important to encourage them to make a conjecture.

Q5 Change the definitions for x and y:

$x(\theta) = \cos(\pi\theta/2)$

$y(\theta) = \sin(\pi\theta/2)$

Q6 Change the sign of any one of the three functions. For example,

$y(\theta) = -\sin(\pi\theta/2)$

(Alternatively, change the signs of all three functions.)

Q7 The radius is the coefficient of the trigonometric functions used for x and y. The height is the range of z. Keeping $[0, 20]$ as the range of θ, these function definitions will render the helix as described:

$x(\theta) = 3\cos(\pi\theta/2)$

$y(\theta) = -3\sin(\pi\theta/2)$

$z(\theta) = 2\theta/5$

THE CONE

Q8 To show both nappes of the cone, enter a negative number for the lower limit.

EXPLORE MORE

Below are rectangular parametric function definitions that will render the sphere and the torus. In both cases, the range of θ is $[0, 10]$.

Sphere:

$$x(\theta) = \sin\left(\frac{\pi\theta}{10}\right)\cos(10\theta)$$

$$y(\theta) = \sin\left(\frac{\pi\theta}{10}\right)\sin(10\theta)$$

$$z(\theta) = \cos\left(\frac{\pi\theta}{10}\right)$$

Torus:

$$x(\theta) = [3 + \cos(\pi\theta)]\cos\left(\frac{39\pi\theta}{5}\right)$$

$$y(\theta) = [3 + \cos(\pi\theta)]\sin\left(\frac{39\pi\theta}{5}\right)$$

$$z(\theta) = \sin(\pi\theta)$$

Representing a line in any coordinate system other than rectangular would be very complicated.

Cylindrical coordinates can greatly simplify the helix, the cone, and the torus.

Spherical coordinates make it easy to bend the curve around a sphere.

VECTOR OPERATIONS IN THREE DIMENSIONS (PAGE 101)

Objective: Students use custom tools to construct the cross product of three-dimensional vectors and use vector operations to determine collinearity and coplanarity of points.

Prerequisites: Students should be familiar with geometric interpretations of the dot product and the cross product.

Sketchpad Proficiency: Advanced. There are many calculations and extensive use of custom tools. Students must be able to perform these tasks independently.

Class Time: 40–50 minutes. The activity is divided into three tasks; you can omit later tasks if time is limited.

Required Sketch: Vector Operations 3D.gsp

Example Sketch: Vector Operations 3D Work.gsp

The sketches used in this activity are perspective representations. Don't use regular Sketchpad measurements (such as **Measure | Angle** and **Measure | Length**). The results would be inconsistent with the geometry being represented. Encourage students to put the axes in motion frequently as they work. They will find it easier to sense the depth of the image when the perspective is changing.

THE CROSS PRODUCT

Q1 The dot products $\mathbf{a} \cdot \mathbf{c}$ and $\mathbf{b} \cdot \mathbf{c}$ should both be zero.

Q2 The cross products $\mathbf{b} \times \mathbf{a}$ and $\mathbf{a} \times \mathbf{b}$ are opposites. They have the same length and opposite directions. Their corresponding coordinates have the same magnitude and opposite signs.

Q3 If the cross product is the zero vector, then vectors **a** and **b** are parallel. Their directions are either the same or opposite, which follows from the formula for the length of $\mathbf{a} \times \mathbf{b}$:

$$|\mathbf{a} \times \mathbf{b}| = |\mathbf{a}| \cdot |\mathbf{b}| \cdot \sin(\angle AOB)$$

If $|\mathbf{a} \times \mathbf{b}|$ is 0, then $m\angle AOB$ must be 0 or π.

COLLINEARITY

Q4 The points are collinear if and only if the cross product $\mathbf{a} \times \mathbf{b}$ is zero. Students may watch the cross product as they change the coordinates. It is easier if they use the custom tool **3D Vector Length** to measure the length of the cross product and watch that instead. If this method is used, an exact solution is unlikely.

Students who are analyzing rather than experimenting may come up with a simple combination of coordinates such that **a** and **b** are scalar multiples of each other, perhaps equal. Some may hit upon an even simpler idea like putting all three points on the x-axis.

COPLANARITY

Q5 The dot product $\mathbf{c} \cdot \mathbf{n}$ is zero if **c** and **n** are perpendicular.

Q6 If **n** is perpendicular to **a, b,** and **c,** then those three vectors lie in a common plane. Points O, A, B, and C lie in that same plane, so they are coplanar. Those four points are translations of P, Q, R, and S, so they too must be coplanar.

One interesting way to confirm this is to construct quadrilateral interior $PQRS$. Drag the perspective controls. If the points are coplanar, it will be possible to find a perspective from which the quadrilateral is viewed on edge, causing it to nearly disappear.

EXPLORE MORE

Q7 This distance is the length of the projection of vector **c** onto **n.** It can be computed using the custom tools provided to compute this value: $|\mathbf{c} \cdot \mathbf{n}|/|\mathbf{n}|$.

Q8 Determining this distance is equivalent to finding the distance from point B to line OA. Let **p** be the projection of vector **b** onto **a**: $|\mathbf{p}| = |\mathbf{b} \cdot \mathbf{a}|/|\mathbf{a}|$. You can derive the displacement, d, from this Pythagorean relationship:

$$d^2 + |\mathbf{p}|^2 = |\mathbf{b}|^2$$

conceptually more difficult method with interesting
:ometry begins with computing another cross product.
et $\mathbf{u} = \mathbf{c} \times \mathbf{a}$. The displacement of point B is the length of
e projection of $\mathbf{b}$ onto $\mathbf{u}$.

$$d = |\mathbf{b} \cdot \mathbf{u}| / |\mathbf{u}|$$

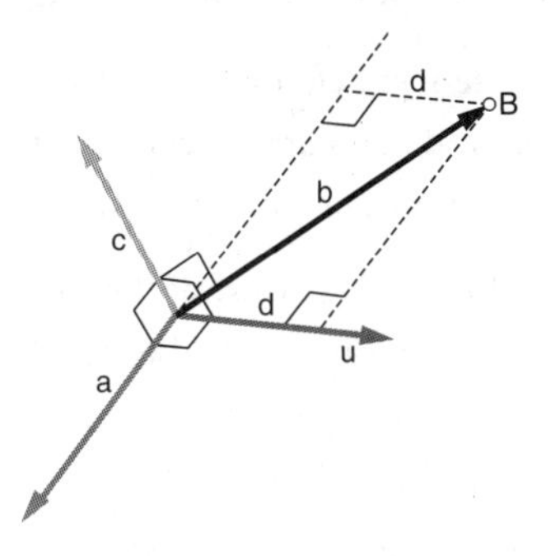

MATRIX TRANSFORMATION IN THREE DIMENSIONS (PAGE 103)

Objective: Students use matrices to transform a shape in three dimensions.

Prerequisites: Students should be familiar with matrix multiplication and with the use of matrices to perform two-dimensional transformations.

Sketchpad Proficiency: Intermediate. Students edit parameters and calculations, and use the custom tools provided. The Explore More section is quite challenging and provides little guidance.

Class Time: 40–50 minutes. There is a lot of editing to do. The questions require only short answers.

Required Sketch: Matrix Transformations 3D.gsp

Example Sketch: Matrix Transform 3D Work.gsp

SIMPLE TRANSFORMATIONS

Q1 The *pitch* control must be at 180°, and the sum of the *spin* and *roll* controls must be 180°.

Q2 Conjectures will vary.

Q3 This is a reflection across the x-z plane.

Q4 When editing the matrix elements, the easiest way to choose the **Edit Parameter** command is from the Context menu, obtained by right-clicking or control-clicking (Mac only) on the parameter. Here are the transformations produced by the matrices:

A: dilate about the origin by scale factor r

B: stretch in the x direction from the y-z plane

C: rotate about the z-axis by angle θ

D: rotate 180° about the y-axis

E: rotate about the y-axis by angle θ

F: dilate about the origin by scale factor -1 (also called a point reflection)

G: project on the x-y plane

H: rotate about the x-axis by angle θ

COMBINED TRANSFORMATIONS

Q5 Point P_1 is in the same position relative to O that P is relative to Q.

Q6 As you change the scale factor, you can see point P_2 dilated by the scale factor about the origin O.

Q7 The path is dilated about point Q rather than the origin.

Q8 The path is rotated about a line through point Q and parallel to the x-axis.

EXPLORE MORE

The described transformation is a rotation about the line $x = z, y = 0$.

MATRIX PRODUCTS IN THREE DIMENSIONS (PAGE 106)

Objective: Students use three-dimensional transformation matrices to simulate the motion of an orbiting planet. This is a three-dimensional extension of the two-dimensional Matrix Product activity.

Prerequisites: Students should first understand the use of 4×4 matrices to effect transformations in three dimensions. They do not have to perform the calculations for matrix products, but they do need to understand this, particularly the concept of working from right to left.

Sketchpad Proficiency: Beginner to intermediate. Students use custom tools and edit parameters to turn them into calculations. No complicated constructions are involved.

Class Time: 30–40 minutes. Most of the work is editing parameters, so the better typists will work much faster. The Explore More challenges are rather difficult and are suitable as a project for advanced students.

Required Sketch: Matrix Product 3D.gsp

Example Sketch: Matrix Product Work 3D.gsp

PLANETARY ORBIT

Show students the completed example sketch so that they have a clear picture of the objective.

An octahedron is used here because it is easy to define the vertex coordinates. Any polyhedron will do, but it should be centered on the origin before transformation.

EXPLORE MORE

See the example file for the sketches described in the challenges. You'll see details about the matrix definitions by double-clicking on the matrix elements.

The moon on page 2 is a cube. It was offset from the origin before transformation. It revolves with the same period as the planet's rotation because the same transformation was applied to both.

The ten-matrix page crowds the screen, but you can get more room by resizing the window or by scrolling down after the matrices have been defined.

On page 3 of the example file is the torus. Quite a few extra angle and scalar controls are used here. To create a new control, copy an existing control along with its corresponding measurement, paste the objects into the same sketch, and change the labels.

Page 4 of the example file is a rotation about line *QR*. Without stereoscopic vision, it is difficult to sense the alignment direction. One thing that can help is to change the view perspective with the *spin* control on the dial. Turn it occasionally between steps, or animate the spin control and leave the animation running.

Matrices **A, B,** and **C** dilate the cube about its center point, *P*. The remaining matrices rotate about line *PQ*. This is a summary of the transformations:

A: translate by vector *PO*

B: dilate by scale factor *r*

C: $= \mathbf{A}^{-1}$

D: translate by vector *QO*

E: rotate about the *z*-axis so that the line rotates into the *z*-*x* plane

F: rotate about the *y*-axis so that the line coincides with the *z*-axis

G: rotate about the *z*-axis by angle *q*

H: $= \mathbf{F}^{-1}$

J: $= \mathbf{E}^{-1}$

K: $= \mathbf{D}^{-1}$

INTRODUCTION TO POLAR COORDINATES (PAGE 111)

Objective: Students experiment with (r, θ) values in polar coordinates by adjusting parameters and find multiple ways of representing the same point. Students then use a right triangle to develop formulas for *x* and *y* in terms of *r* and θ.

Prerequisites: Students must be familiar with the definition of sine and cosine in a right triangle.

Sketchpad Proficiency: Intermediate. Students start with a blank sketch in which they create a polar coordinate system, create parameters, plot points, measure coordinates, and use the Calculator. Each command students must use is explicitly described in the directions.

Class Time: 25–35 minutes

Required Sketch: None

Example Sketch: Intro To Polar.gsp

SKETCH AND INVESTIGATE

1. For this activity to work correctly, the angle units must be set to directed degrees.

Q1 Conjectures will vary; it is only important that students make a prediction of some kind.

Q2 The value of *r* determines how far the plotted point is from the pole. The value of θ determines the angle the ray from the pole through the point makes with the positive horizontal axis.

Q3 Point *A* lands on the target for values such as (3, 510°), (3, 870°), and (3, −210°). The value of *theta* must be of the form $150° + k \cdot 360°$, where *k* is any integer.

Q4 When *theta* is 150°, there is no value of *r* other than 3 that will make *A* land on the target.

Q5 Point *A* lands on the target for values such as (−3, 330°), (−3, 690°), and (−3, −30°). The value of *r* must be −3, and the value of *theta* must be of the form $330° + k \cdot 360°$, where *k* is any integer.

CONVERT BETWEEN THE TWO FORMS

Q6 The value of x is given by $x = r\cos\theta$.

Q7 $y = r\sin\theta$

Q8 $r = \pm\sqrt{x^2 + y^2}$. The $\pm$ sign is necessary because specific values for x and y do not uniquely determine a single value for r, as students discovered in Q5.

Q9 The short answer is $\theta = \tan^{-1}\left(\frac{y}{x}\right)$. However, this does not tell the whole story, because this formula always gives a result such that $-90° < \theta \leq 90°$. In fact there are multiple values of θ for any given values of x and y, as students discovered in Q3 and Q5, so a more complete answer is $\theta = \tan^{-1}\left(\frac{y}{x}\right) + k \cdot 180°$, where k is any integer. Half of these values correspond to positive values of r, and half to negative values of r.

EXPLORE MORE

Q10 The graph of $r = 6$ is a circle centered at the pole, with a radius of 6 units. The function $r = -6$ produces the same graph.

Q11 The graph of $\theta = 75°$ is a line through the pole, making an angle of 75° with the positive x-axis. Identical graphs are produced by $\theta = 255°$, $\theta = 435°$, and any function of the form $\theta = 75° + k \cdot 180°$, where k is any integer.

Q12 The equation $x = 6$ is equivalent to $r\cos\theta = 6$, so the polar function is

$$r = \frac{6}{\cos\theta}, \text{ or } r = 6\sec\theta$$

Similarly, $y = -2$ is equivalent to $r\sin\theta = -2$, so the polar function is

$$r = \frac{-2}{\sin\theta}, \text{ or } r = -2\csc\theta$$

Q13 The function is $r = \theta$.

CARTESIAN GRAPHS AND POLAR GRAPHS (PAGE 113)

Objective: Students compare rectangular graphs and polar graphs for functions in the form $y = a\sin(bx)$ and $r = a\sin(b\theta)$. Students analyze how the period and amplitude of a Cartesian graph correlate with features of the corresponding polar graph.

Prerequisites: Students should be familiar with the graphs of sinusoids, and specifically with the concepts of period and amplitude. This activity is best utilized shortly after students have been introduced to graphing in polar coordinates.

Sketchpad Proficiency: Beginner. Students manipulate a pre-made sketch.

Class Time: 20–30 minutes

Required Sketch: Cartesian Polar.gsp

SKETCH AND INVESTIGATE

Q1

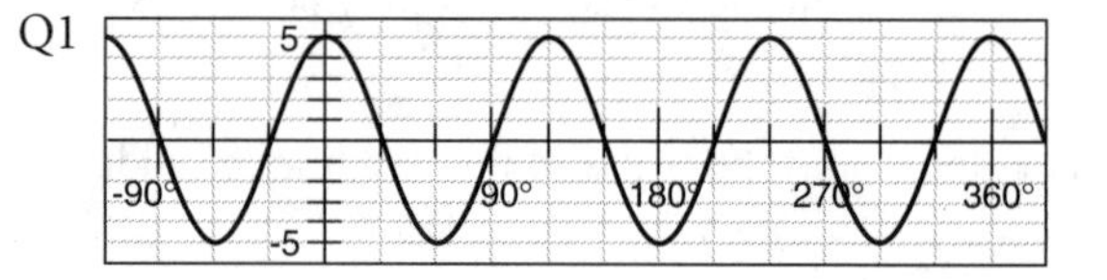

The parameter a controls the amplitude of the sinusoid. The parameter b controls the period.

Q2 Answers will vary. The point of this question is to get students thinking about the process before they reveal the answer. Students have no basis yet for guessing correctly, so emphasize that the important thing is guessing, not getting the answer right.

Q3 The x-intercepts on the Cartesian graph correspond to $r = 0$, where the curve crosses the pole (origin) of the polar graph.

Q4 The maximum and minimum points on the Cartesian graph correspond to the outermost point of each leaf of the polar graph. The only difference between the two types of points on the polar graph is that the maximum points are created when the output value (r) is positive (the "bowtie" is on the positive side of the output bar), and the minimum points are created when the output value is negative (the "bowtie" is on the negative side of the output bar).

Q5 The polar graph starts on the positive side of the output axis for the first repetition (starting at 0°) and starts on the negative side of the output axis for the second repetition (starting at 180°). At all corresponding points (points whose θ-values are separated by 180°), r-values are opposite.

Q6 The parameter a controls the distance from the pole to the tip of each leaf.

Q7 When the parameter b is odd, the number of leaves is equal to b. When b is even, the number of leaves is equal to $2b$.

XPLORE MORE

Q8 Between 0° and 360° (not including 360°), the graph of $y = a\cos(bx)$ has b maximums and b minimums for a total of $2b$ extreme points—the points that become the outer points on the leaves. For even integer values of the parameter b, the Cartesian graph yields a maximum y-value at $\theta = 0°$ and $\theta = 180°$, meaning that the graph doesn't double back on itself and there will be the maximum ($2b$) number of leaves. For odd integer values of the parameter b, the Cartesian graph yields a maximum y-value at $\theta = 0°$, but it yields a minimum at $\theta = 180°$, causing the graph to double over itself at 180° and resulting in only b leaves.

Q9 This equation produces a vertical line 2 units to the right of the pole. Students can use trigonometry to explain why this occurs. Consider the right triangle in the following diagram, using an arbitrary point B on the vertical line through point A at (2, 0°).

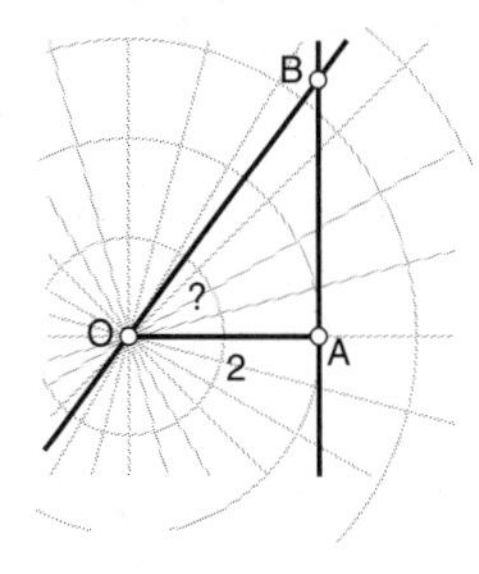

To find the length of the hypotenuse OB, use the cosine function:

$$\cos(\theta) = \frac{OA}{OB} = \frac{2}{OB}$$

$$OB = \frac{2}{\cos(\theta)} = 2\sec(\theta)$$

Thus the equation of any point on the vertical line must be $r = 2\sec(\theta)$.

Q10 Three interesting functions to try are

$$f(\theta) = a\tan(b\theta),$$

$$f(\theta) = a\left(\frac{\theta}{90°}\right),$$

$$\text{and } f(\theta) = a\sqrt{\frac{\theta}{90°}}$$

MULTIPLICATION OF COMPLEX NUMBERS
(PAGE 115)

Objective: Students use a set of Sketchpad custom tools to build, step by step, the product of two complex numbers. By analyzing their construction, students develop a geometric understanding of complex number multiplication.

Prerequisites: Students should be familiar with the algebra of complex numbers as well as the SAS triangle similarity theorem.

Sketchpad Proficiency: Beginner. Students use pre-built custom tools.

Class Time: 40–50 minutes

Required Sketch: Complex Multiplication.gsp

Presentation Sketch: Complex Multiplication 2.gsp is a pre-built model that allows students to multiply any two complex numbers (represented as vectors). The lengths and arguments of the vectors are calculated automatically.

This activity is based on an idea of Bowen Kerins and the High School Teachers Program Group of the Park City Mathematics Institute.

SKETCH AND INVESTIGATE

Q1 The vector $\overrightarrow{OF}$ represents **wz.**

Q2 The length of $\overrightarrow{OC}$ is $a|\mathbf{z}|$.

Q3 The length of $\overrightarrow{OD}$ is $|\mathbf{z}|$, and the length of $\overrightarrow{OE}$ is $b|\mathbf{z}|$.

Q4 The vector $\overrightarrow{CF}$ is a translation of $\overrightarrow{OE}$, so its length is the same: $b|\mathbf{z}|$.

Q5 Vector $\overrightarrow{OE}$ was obtained by multiplying $\overrightarrow{OZ}$ by i and then dilating it by b. Multiplication by i corresponds to a rotation of 90°. Thus $\overrightarrow{CF}$, a translation of $\overrightarrow{OE}$, makes a 90° angle with $\overrightarrow{OC}$.

Q6 Both $\triangle OVW$ and $\triangle OCF$ are right triangles. The sides OC and CF of $\triangle OCF$ are each $|\mathbf{z}|$ times as long as the sides OV and VW of $\triangle OVW$. Thus, by the SAS triangle similarity theorem, the two triangles are similar.

Q7 The scale factor of the two similar triangles is $|\mathbf{z}|$. Thus the length of $\overrightarrow{OF}$ is $|\mathbf{w}| \cdot |\mathbf{z}|$.

Q8 Because $\triangle OVW$ and $\triangle OCF$ are similar, both $\angle WOV$ and $\angle COF$ equal α. Thus the argument of $\overrightarrow{OF}$ is $\alpha + \beta$

Q9 To multiply two complex numbers w and z, *multiply* their lengths and *add* their arguments.

Q10 The length of **wz** is 6, and its argument is 150°. Drawing a picture and applying the relationships found in a 30-60-90 triangle yields $\mathbf{wz} = -3\sqrt{3} + 3i$

EXPLORE MORE

Q11 The complex number with length 1 and argument 270° equals -1 when squared. This number is $-i$.

Q12 The complex number with length 1 and argument 45° will equal i when squared. Written in the form $x + iy$, this number is $\frac{\sqrt{2}}{2} + \frac{i\sqrt{2}}{2}$. Similarly, a complex number with length 1 and argument 225° will also equal i when squared. This number is $\frac{-\sqrt{2}}{2} - \frac{i\sqrt{2}}{2}$.

Q13 The complex numbers $a + bi$ and $a - bi$ are reflections of each other across the real axis. Their arguments sum to 0° (equivalently, 360°) when the numbers are multiplied. Thus the product sits on the positive real axis.

I SEARCH OF BURIED TREASURE (PAGE 118)

his activity is based on an engaging problem from the book *ne Two Three . . . Infinity,* by the physicist George Gamow.) find a fictional buried treasure, students use both gebraic and geometric properties of the complex plane.

bjective: Students construct a sketch that models the rections given to unearth a buried treasure. They make a njecture about the treasure's location and then prove their sults by overlaying the complex plane onto their map.

rerequisites: Students should be familiar with the mplex plane and the algebra of finding a segment's idpoint given the coordinates of its endpoints. They also ould know that when a complex number is rotated 90° out the origin, this corresponds to multiplication by i.

etchpad Proficiency: Intermediate. Students build a etch from scratch that involves rotation and the nstruction of a segment's midpoint.

lass Time: 40–50 minutes

equired Sketch: Buried Treasure.gsp

KETCH AND INVESTIGATE

efore turning to Sketchpad, you can ask students to model e treasure directions on your classroom floor or in the hoolyard. Mark the location of the two trees and pick a ndom spot for the scarecrow. Use string or pace out steps find and mark the treasure. Do so again for a new cation of the scarecrow. Students should discover that e treasure is in roughly the same location in both cases.

Q1 The treasure, point T, stays in the same spot regardless of the location of point S. Surprisingly, then, it doesn't matter where the scarecrow once stood. The walking directions lead to the same spot for every possible scarecrow location you might choose.

ROVE YOUR RESULTS

Q2 Point T is at i on the complex plane. Two other noteworthy observations are

- The origin is equidistant from the oak tree, the elm tree, and the treasure.
- The treasure is on the perpendicular bisector of the segment connecting the oak and elm trees.

Q3 Moving the origin 1 unit to the left shifts the real component of a complex number by $+1$. Thus the new location of the scarecrow is $(a + 1) + bi$.

Q4 Multiplication by i corresponds to a rotation of 90°.

Q5 The location of M_1 is iS, which equals $-b + i(a + 1)$.

Q6 Moving the origin 1 unit back to the right shifts the real component of a complex number by -1. Thus the new location of M_1 is $-b - 1 + i(a + 1)$.

Q7 The new location of the scarecrow is $(a - 1) + bi$.

Q8 Multiplication by $-i$ corresponds to a rotation of $-90°$.

Q9 The location of M_2 is $-iS$, which equals $b - i(a - 1)$.

Q10 The new location of M_2 is $b + 1 - i(a - 1)$.

Q11 To find the midpoint of M_1M_2, add M_1 to M_2 and then compute the average of the real and imaginary components. $M_1 + M_2 = [-b - 1 + i(a + 1)] + [b + 1 - i(a - 1)] = 0 + 2i$. Dividing by 2 yields i. Thus, as seen in the sketch, the treasure sits at i.

PROJECT IDEAS

Other proofs of the treasure problem also exist. Ask students to research the proofs found in the February 1998 and September 2003 issues of *Mathematics Teacher* magazine.

TRANSFORMATIONS IN THE COMPLEX PLANE (PAGE 120)

Objective: Students explore a collection of sketches, each of which displays a complex number, z, and, simultaneously, another complex number, w. The mystery number w is computed as a function of z. By dragging z and watching the behavior of w, students match each w to an equation listed in the activity.

Prerequisites: Students should be familiar with the algebra of complex numbers.

Sketchpad Proficiency: Beginner. Students work with pre-made sketches.

Class Time: 30–40 minutes

Required Sketches: Complex Transformations.gsp and **Complex Transformations 2.gsp**

SKETCH AND INVESTIGATE

Q1 The list below indicates which page of the Sketchpad file corresponds to the given transformation.

a. page 6

b. page 10

c. page 2

d. page 8

e. page 3

f. page 9

g. page 7

h. page 4

i. page 1

j. page 5

Q2 Students can use what they observed in Q1 to help predict the transformations of the letter *F*. On page 1 of the sketch, for example, multiplication by i corresponds to a rotation of 90° about the origin. Thus the locus of w will be a rotation of the letter *F* by 90°.

POWERS OF COMPLEX NUMBERS (PAGE 121)

Objective: By exploring several interactive models, students discover that multiple values of z can satisfy an equation like $z^3 = 2i$. As students explore higher and higher powers of z, they begin to see patterns that form the basis of De Moivre's Theorem.

Prerequisites: Students should be familiar with the geometric meaning of complex number multiplication. (The activity Multiplication of Complex Numbers is a good lesson to do first.) They should also be familiar with the calculation of sine and cosine values.

Sketchpad Proficiency: Beginner. Students work with a pre-made sketch.

Class Time: 40–50 minutes

Required Sketch: Complex Powers.gsp

Presentation Sketch: Complex Powers.gsp

PRELIMINARY WORK

Q1 To multiply two complex numbers, multiply their lengths and add their arguments.

Q2 To square a complex number, square its length and double its argument.

Q3 $2i = 2(\cos 90 + i\sin 90)$

Q4 Based on Q2 and Q3, $z = \sqrt{2}(\cos 45 + i\sin 45)$ is equal to $2i$ when squared.

INVESTIGATE z^2

Q5 $z = 1 + i$

To position z as precisely as possible (by a tenth of a pixel at a time), students can use the movement buttons.

Note: For most of the remaining Sketchpad-related questions, even moving by a tenth of a pixel at a time does not allow students to obtain an exact answer. You can ask students how to calculate precise answers after they obtain their Sketchpad approximations. (Trigonometry is of great help.)

Q6 The complex number $-(1 + i)$ also equals $2i$ when squared since $[-(1 + i)]^2 = (-1)^2 \cdot (1 + i)^2 = 1 \cdot 2i$.

If students get stuck on this question, encourage them to drag z around while looking for different positions that place z^2 at $2i$.

Q7 $z = \sqrt{2}(\cos 225 + i \sin 225)$

Q8 Squaring the value of z from Q7 yields $z^2 = 2(\cos 450 + i \sin 450) = 2i$.

HIGHER POWERS OF z

Q9 Written in the form $a + bi$, the three values of z are $\sqrt[3]{2}(\sqrt{3}/2 + 1/2i)$, $\sqrt[3]{2}(-\sqrt{3}/2 + 1/2i)$, and $-\sqrt[3]{2}i$. Sketchpad will express these values in decimal notation.

Written in the form $r(\cos\theta + i\sin\theta)$, the three values of z are $\sqrt[3]{2}(\cos 30 + i\sin 30)$, $\sqrt[3]{2}(\cos 150 + i\sin 150)$, and $\sqrt[3]{2}(\cos 270 + i\sin 270)$.

Q10 To cube a complex number z, cube its length and triple its argument.

Q11 Cubing $\sqrt[3]{2}$ and tripling the angle measures of the values from Q9 yields three complex numbers that equal $2i$.

Q12 The four values of z are $\sqrt[4]{2}(\cos 22.5 + i\sin 22.5)$, $\sqrt[4]{2}(\cos 112.5 + i\sin 112.5)$, $\sqrt[4]{2}(\cos 202.5 + i\sin 202.5)$, and $\sqrt[4]{2}(\cos 292.5 + i\sin 292.5)$.

Q13 The five values of z are $\sqrt[5]{2}(\cos 18 + i\sin 18)$, $\sqrt[5]{2}(\cos 90 + i\sin 90)$, $\sqrt[5]{2}(\cos 162 + i\sin 162)$, $\sqrt[5]{2}(\cos 234 + i\sin 234)$, and $\sqrt[5]{2}(\cos 306 + i\sin 306)$.

Q14 On page 1, the two red x's form a segment passing through the origin. On page 2, three red x's form the vertices of an equilateral triangle. On page 3, the x's form a square, and on page 4, the x's form a regular pentagon.

EXPLORE MORE

Q15 In addition to using Sketchpad, students can apply their results from previous questions to solve $z^n = -1$. The argument of one z-value satisfying the equality is $180/n$. You can find the remaining $n - 1$ arguments by adding $360/n$ degrees repeatedly to $180/n$. This pattern forms the basis of De Moivre's Theorem.

A GEOMETRIC APPROACH TO $e^{i\pi}$ (PAGE 123)

Douglas Hofstadter, author of *Gödel, Escher, Bach,* wrote that when he first saw the statement $e^{i\pi} = -1$, "... perhaps at age 12 or so, it seemed truly magical, almost otherworldly." The goal of this activity is to bring this well-known statement down to earth by providing an informal geometric interpretation of its meaning.

Objective: Students use a limit definition of e, along with multiplication on the complex plane, to find the value of $e^{i\pi}$.

Prerequisites: Students should first complete the Multiplication of Complex Numbers activity. They should also have at least some exposure to the concept of limits.

Sketchpad Proficiency: Beginner. Students use Sketchpad's Calculator and follow step-by-step instructions for creating iterations.

Class Time: 40–50 minutes

Required Sketch: eipi.gsp

This activity is based on an idea of John Conway and Richard Guy.

GETTING STARTED

Q1 The four calculations seem to approach a value roughly equal to 2.71.

Q2 A good approximation is 2.71828.

Q3 A good approximation of e^3 is 20.0855.

SKETCH AND INVESTIGATE

Q4 To multiply $1 + i\pi/10$ by itself, square its length (OB) and double its argument (AOB). To compute the tenth power of $1 + i\pi/10$, raise its length to the tenth power and multiply its argument by 10.

Q5 The successive powers of $1 + i\pi/10$ are represented in counterclockwise order by the vectors that originate at point O.

Q6 The vector OP represents $(1 + i\pi/10)^{10}$. The coordinates of point P are $(-1.59, 0.16)$. Thus, $(1 + i\pi/10)^{10}$ equals $-1.59 + 0.16i$.

Q7 The value of $(1 + i\pi/n)^n$ approaches -1 as n grows larger.

EXPLORE MORE

Q8 The value of $e^{i\pi/2}$ is i.

Q9 The value of $e^{i\pi/3}$ is approximately $0.5 + 0.87i\left(\cos\frac{\pi}{3} + i\sin\frac{\pi}{3}\right)$ and the value of $e^{i\pi/4}$ is approximately $0.71 + 0.71i\left(\cos\frac{\pi}{4} + i\sin\frac{\pi}{4}\right)$.

Q10 As the approximations get better and better, the length of the vector approaches 1.

Q11 When $k = 2$, the angle is $\frac{\pi}{2}$. When $k = 3$, the angle is $\frac{\pi}{3}$. When $k = 4$, the angle is $\frac{\pi}{4}$.

Q12 The vector representing $e^{i\theta}$ has a length of 1 and makes an angle of θ with the x-axis. Thus, its real component is $\cos\theta$ and its complex component is $\sin\theta$.

Q13 When $\theta = \pi$, $e^{i\pi} = \cos\pi + i\sin\pi = -1$.

GENERATING ARITHMETIC AND GEOMETRIC SEQUENCES NUMERICALLY (PAGE 129)

Objective: Students develop an understanding of arithmetic and geometric sequences by building and modifying them with Sketchpad.

Prerequisites: Students should be familiar with the definitions of an arithmetic and geometric sequence.

Sketchpad Proficiency: Beginner. Students work with a pre-made sketch and follow step-by-step instructions to perform a numerical iteration.

Class Time: 25–35 minutes

Required Sketch: Sequences.gsp

This activity is based on an idea and sketch by Nathalie Sinclair.

ARITHMETIC SEQUENCES

Q1 The number line shows the arithmetic sequence 2, 5, 8, 11, 14, ... The *start* value is the first term in the sequence, and the *difference* value is the difference between any two consecutive terms.

Q2 There are many ways to change the sequence so that it includes 24. Changing the *start* value to 3 or the *difference* value to 2 both work.

Q3 a. *start* = 0; *difference* = 3

b. *start* = −6; *difference* = −4

c. *start* = 1; *difference* = 0

d. *start* = 0.25; *difference* = 0.25

Q4 There are nine numbers between 2000 and 2010, and the difference between terms in the sequence is only 6. Thus there must be a term in the sequence between 2000 and 2010.

GEOMETRIC SEQUENCES

Q5 Two possibilities are: *start* = 3; *ratio* = 2 and *start* = 8; *ratio* = 3.

Q6 a. *start* = 0.5; *ratio* = 4

b. *start* = −64; *ratio* = −0.5

c. *start* = 1; *ratio* = 1

d. *start* = −1; *ratio* = −1

Q7 Three copies of the 2nd arc fit into the 3rd arc, and three copies of the 3rd arc fit into the 4th arc. This pattern continues. Students can use algebra to prove a more general result for any geometric sequence.

XPLORE MORE

Q8 To generate a Fibonacci sequence, map *seed1* to *seed2* and *seed2* to *seed1* + *seed2*.

AREA MODELS OF GEOMETRIC SERIES
(PAGE 131)

Objective: Students use dissections of a square to represent geometric series and to investigate sums of these series.

Prerequisites: Students should be familiar with the definition of an infinite geometric series.

Sketchpad Proficiency: Advanced. Students use custom tools and iteration.

Class Time: 40–50 minutes

Required Sketch: Area Models.gsp

Example Sketch: Area Models Work.gsp

DISSECTION 1

Q1 Quadrilateral *EBCF* represents 1/2 the area of the entire quadrilateral *ABCD*.

Q2 The iterated quadrilaterals will fill the entire original figure, so the sum of the series is 1.

Q3 When the depth is 17, the partial sum is greater than 0.99999 and shows as 1.00000 in the table.

Q4 The partial sum will never be exactly 1, no matter how many iterations are done. The value of 1 is the limit, which means that the partial sum gets closer and closer to 1 the more iterations there are, but never quite reaches 1. If there could be an infinite number of iterations, the partial sum would be 1.

DISSECTION 2

Q5 This polygon represents 1/3 of the total area, because it includes three of the nine smaller squares.

Q6 The three new polygons each represent 1/3 of one of the smaller squares, each of which is 1/9 of the large square. The area of one of the new polygons is thus $1/3 \cdot 1/9$, and the area of all three together is 1/9.

Q7 The polygons appear to fill half the large square. Thus the sum of the geometric series is 1/2.

DISSECTION 3

Q8 Each of the polygons represents 1/3 of the area of the large square.

Q9 When you increase the depth, the two spirals intertwine, filling the entire large square, so each spiral represents 1/2 of the large square. The partial sum of the series is thus approaching 1/2.

Q10 When the depth of iteration is 10, the partial sum exceeds 0.49999.

EXPLORE MORE

Q11 The shapes can be arbitrary parallelograms. If you change the original shape in such a way that its area changes, you must click the *Calibrate Area* button so that the relative area measurement is accurate.

Q12 Each piece is 1/5 of the total area. Here's one of the outside pieces:

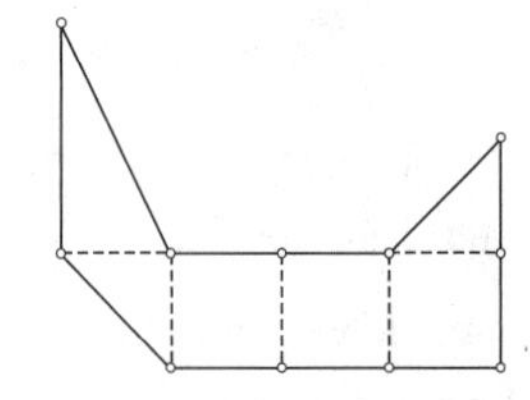

Each of the three squares represents 1/25 of the total. The two small triangles combine to make another square, and the larger triangle (with a base of 1 and height of 2) is also equivalent to another square. So the polygon's area is 5/25 of the total.

A similar dissection of the center square shows that it is 1/5 of the total.

Q13 Because the four iterated spirals fill the original figure, each spiral has an area of 1/4. Each spiral starts with a piece that has an area of 1/5, and then adds a similar area inside a square that's 1/5 the size of the original, so the second piece has an area of 1/25. Similarly, the area of the third piece of each spiral is 1/125, and so on. This figure represents the series

$$\frac{1}{5}+\frac{1}{25}+\frac{1}{125}+\cdots=\frac{1}{4}$$

Q14 Each piece of this square is 1/4 of the total. This is easy to see for the square in the lower right corner. For the trapezoid in the lower left, the height is 1/2 and the two bases are 2/3 and 1/3. Thus the area is

$$\frac{(b_1+b_2)}{2}\cdot h=\frac{\left(\frac{2}{3}+\frac{1}{3}\right)}{2}\cdot\frac{1}{2}=\frac{1}{4}$$

The area of the upper right trapezoid is identical. Th remaining piece can be found by subtracting the known pieces from 1, or by constructing segment *CI* to make it into two trapezoids and finding the area o the trapezoids.

Q15 Each smaller shape is 1/4 of the total area. The iterations of these shapes fill the entire figure, so each iteration has an area of 1/3. Therefore this figure shows the geometric series

$$\frac{1}{4}+\frac{1}{16}+\frac{1}{64}+\cdots=\frac{1}{3}$$

Q16 When n is an integer and $n \geq 2$, the sum will always be $1/(n-1)$. This relation can be seen by any dissection of the large square into n smaller areas. Color $n-1$ of them, and then subdivide the one remaining small area ($1/n$ of the total) into n still smaller areas. Each of these smaller areas represents $1/n^2$ of the total—the 2nd term of the series. Again color $n-1$ of them using the same colors as before, and repeat the process indefinitely. The result will be a square colored with $n-1$ colors, each of which represents a region (not necessarily connected) with area $1/(n-1)$. So the sum of all the regions with this same color—that is, the sum of the geometric series—is $1/(n-1)$.

GEOMETRIC SERIES COIL (PAGE 134)

is activity offers a visual model of a geometric series. sing Sketchpad, students build an angular "coil" mposed of segments whose lengths form a geometric ries. By adding more and more segments to the coil and winding it flat, students are able to find the limiting sum the geometric series.

bjective: Through a visual and numerical exploration, idents see firsthand what happens to a geometric series ien the number of its terms grows larger and larger.

erequisites: None

etchpad Proficiency: Beginner or intermediate, pending on whether students follow the general structions or the detailed instructions.

ass Time: 40–50 minutes

quired Sketch: Geometric Coil.gsp

ample Sketch: Geometric Coil Work.gsp

JM THE SERIES

1 The next three lengths are 0.75^2, 0.75^3, and 0.75^4. The infinite series represented by the coil is $1 + 0.75 + 0.75^2 + 0.75^3 + 0.75^4 + \ldots$

2 The sum depends on the number of iterations.

3 The sum of the segments would approach 4. As a result of rounding, Sketchpad displays the sum as 4 when the coil grows to 48 segments in length. Students can display the sum to greater precision by changing the Value Properties of the sum. When calculated precisely, the sum gets infinitely close to 4, but never equals 4. This is an important point to discuss with your students.

EATE A TABLE OF VALUES

4 The series grows beyond 3.99 when $n = 20$. This corresponds to the sum $1 + 0.75 + 0.75^2 + \ldots + 0.75^{20}$.

5 The sum of the series is $= 1/(1 - \mathit{ratio})$.

6 When the ratio is greater than or equal to 1, the coil grows larger and larger and its sum does not converge to a finite value.

A GEOMETRIC SERIES STAIRCASE (PAGE 137)

When summing an infinite geometric series, precalculus texts offer a standard technique: multiply the series by its common ratio r and then subtract the new identity from the original statement. As a companion to the traditional derivation, this activity offers a novel summation method that is more visual in its approach.

Objective: Students build a model depicting a staircase of shrinking squares. They examine the model to find a segment whose length represents the sum of an infinite geometric series. Through the application of similar triangles, students derive the generalized algebraic formula for the sum of an infinite geometric series.

Prerequisites: Students should be familiar with the definition of an infinite geometric series (see the A Geometric Series Coil activity). The proof involves an application of similar triangles.

Sketchpad Proficiency: Intermediate. Students follow step-by-step instructions that involve dilation, rotation, iteration, and perpendicular line construction.

Class Time: 40–50 minutes

Required Sketch: Geometric Staircase.gsp

Example Sketch: Staircase Work.gsp

This activity is based on an idea of Ray Hemmings and Dick Tahta.

CREATE A STAIRCASE

1. Students can mark the center and ratio of dilation ahead of time, by using the **Mark** commands in the Transform menu. But it's easier to follow the directions here, selecting the square and choosing the **Dilate** command first, and then clicking in the sketch on the ratio to use.

SUM THE SERIES

Q1 The side lengths of the next four squares are ar^2, ar^3, ar^4, and ar^5. These are terms of a geometric sequence with first term a and common ratio r.

Q2 The sum of the squares' side lengths is equal to S (assuming you could iterate the squares an infinite number of times). If you imagine taking all the squares and pushing them flush against either $\overline{JE}$ or $\overline{JI}$, you can see that segments JE and JI are both equal to S.

Q3 $\angle CBF$ and $\angle BIJ$ are both complementary to $\angle IBJ$. Thus, the two angles are congruent, and triangles BCF and IJB are similar.

Q4 Based on the similar triangles,

$$\frac{CF}{BC} = \frac{JB}{IJ}$$

In terms of S, r, and a, the proportion becomes

$$\frac{ar}{a} = \frac{JB}{S}$$

Thus, $JB = Sr$.

Q5 Rewriting $JB + BE = JE$ gives $Sr + a = S$. Solving for S yields

$$S = \frac{a}{1 - r}$$

Q6 When r is greater than or equal to 1, the squares do not shrink in size. The squares form a geometric sequence of increasing terms whose sum does not converge to a finite value. In geometric terms, the intersection (point I in the sketch) no longer exists, at least not in the direction in which the series is represented.

TAYLOR SERIES (PAGE 139)

Objective: Students create a Taylor series and explore how adding terms to a Taylor series approximation increases the accuracy of the approximation.

Prerequisites: Students must be familiar with the sine and cosine functions and ideally should have some introduction to what a Taylor series is. Although students work with a Taylor series in this activity, it doesn't attempt to explain the basis of Taylor series.

Sketchpad Proficiency: Intermediate

Class Time: 30–40 minutes

Required Sketch: None

Presentation Sketch: Taylor Series Present.gsp

SKETCH AND INVESTIGATE

Q1 To calculate the next value of i, calculate $i + 2$.

Q2 For the next value of *num*, calculate $-num \cdot x^2$.

Q3 To calculate the next value of *den*, calculate $den \cdot (i + 1)(i + 2)$.

Q4 The initial value of the sum, before adding the first term, should be zero. To calculate the next value of *sum*, calculate $sum + num/den$.

Q5 The initial values should be $i = 1$, $den = 1$, and $sum = 0$. Because *num* depends on x, set it to 2.

Q6 The locus is a straight line—the graph of $f(x) = x$. Only the first term of the series contributes to this graph.

Q7 Now the shape is that of a cubic function, generated by the first two terms: $f(x) = x - x^3/6$.

Q8 The terms are polynomials of order 1, 3, 5, 7, and so forth. The high-order terms have large denominators so the low-order terms predominate for small values of x; the high-order terms introduce corrections for larger values of x.

Q9 About 8 terms gives a reasonable approximation for a single period; about 16 terms are required for two periods.

The Taylor series expansion of the sine function doesn't converge rapidly enough to be of use in calculators and in computer algorithms. Other, faster-converging series are used for these purposes.

If you did want to use the Taylor series for these calculations, you'd get the best results by first transforming the argument (the x-value) so that it is between 0 and $\pi/2$ and then performing the calculation. Ask students how they would do this—they should figure out how to do the transformation, and they should describe how to keep track of whether the resulting value should be positive or negative.

A related follow-up exercise would be to ask students to use Sketchpad's calculator to determine the number of digits of accuracy they obtain for sin $(\pi/2)$ for various depths of iteration. As the following Sketchpad-generated table shows, eight iterations gives a result accurate to ten decimal places.

terms	value	error $\cdot$ 10^6
1	1.57080	570796.32679
2	0.92483	−75167.77071
3	1.00452	4524.85553
4	0.99984	−156.89860
5	1.00000	3.54258
6	1.00000	−0.05626
7	1.00000	0.00066
8	1.00000	−0.00001
9	1.00000	0.00000

PRESENT

A nice touch, when presenting this activity, would be to create a fifth parameter, *depth*, and to do the iteration to a depth determined by this parameter. Then you can animate the resulting iteration with a button that changes the *depth* parameter.

EXPLORE MORE

To generate the Taylor series approximating the cosine function, you need only change the initial value of i to 0 and the initial value of *num* to 1.

DEMONSTRATE

You can use the sketch **Taylor Series Present.gsp** in a classroom demonstration, using the buttons to move through the demonstration.

INSTANTANEOUS RATE (PAGE 145)

Objective: Students make a connection between instantaneous rate and slope of the tangent to the graph, see the instantaneous rate as a limit of the slope between two points (just as the tangent represents the limit of a secant line), and are introduced to the concept and definition of the derivative.

Prerequisites: Students should be familiar with finding the slope given two points on a graph.

Sketchpad Proficiency: Beginner. Students work with a pre-made sketch.

Class Time: 30–40 minutes

Required Sketch: Instantaneous Rate.gsp

DOOR ANGLE AS A FUNCTION OF TIME

Q1 The angle increases from $t = 0$ to approximately $t = 1.45$. You can tell because the maximum value of the angle occurs at approximately $t = 1.45$.

Q2 The maximum angle is approximately 106°, at $t = 1.45$.

THE DOOR AT TWO DIFFERENT TIMES

Q3 The slider has a maximum value of 1. The smallest separation you can observe on the graph is about 0.01. The slider uses a logarithmic scale, so it's easy to achieve very small values—values that are much smaller than you can observe on the graph.

Q4 Yes, the displayed values (to five decimal digits) continue to show the changes.

THE RATE OF CHANGE OF THE DOOR'S ANGLE

Q5 The units are degrees/s. The rate of change tells you by how many degrees the door is opening or closing for every second.

Q6 The rate of change is the slope of the dotted secant line. As the secant line approaches tangency, the calculated rate of change approaches the instantaneous rate of change.

Q7 When the rate of change is positive, the door is opening; when the rate is negative, the door is closing. When the rate of change is close to zero, the door is moving slowly; when the absolute value of the rate of change is large, the door is moving quickly.

Q8 When $t_1 = 1$ and $\Delta t = 0.1$, the calculated rate of change is 26.33629 degrees/s.

THE LIMIT OF THE RATE OF CHANGE

Q9 The average rate of change is higher (more than 30 degrees/s) and appears to be a more accurate value for $t_1 = 1.0$.

Q10 The rate of change appears to approach a limit; the differences are less and less. The rate of change seems to be approaching 30.685 degrees/s.

Q11 No, the change is too small to observe on the graph.

Q12 The derivative (that is, the limit of the rate of change) when t_1 is 3 seconds seems to be about -26.986 degrees/s. The negative sign means that the door is closing.

ONE TYPE OF INTEGRAL (PAGE 147)

Objective: Students explore the concept of definite integral—the accumulation of a function's values over a particular domain of the independent variable. The practical application, in which students use the velocity function to determine distance traveled, helps to suggest the usefulness of the definite integral.

This activity emphasizes the nature of the definite integral as the accumulation of the value of a function over a particular domain. Counting squares makes the "accumulation of value" concrete and can also leave students motivated to learn the mathematics that will enable other less tedious methods.

Prerequisites: Students should understand the relationship between rate, time, and distance traveled. You'll need to spend some time interpreting the graph if students aren't already familiar with graphs of velocity as a function of time.

Sketchpad Proficiency: Beginner. No construction is required. Students will need to double-click parameters to change their values.

Class Time: 40–50 minutes

Required Sketch: Definite Integral.gsp

THE DISTANCE A CAR TRAVELS

This activity uses the term *velocity* rather than *speed,* because direction is important when accumulating the values. If the driver puts the car in reverse, the velocity is negative, and the distance traveled is reduced—but the speed (which is the absolute value of the velocity) is positive, even when the car is traveling backward.

1. Some students may expect that the velocity of the point on the screen corresponds to the velocity of the car. It's important for students to realize that the car's velocity corresponds to the vertical position on the graph, and *not* to the velocity of the point on the graph. It may be useful to have the class observe the animation and interpret the graph as a group.

Q1 Each grid square is 5 s wide.

Q2 Each grid square is 5 ft/s high.

Q3 Each grid square represents 25 ft of travel.

Q4 The region on the right has $6 \cdot 12 = 72$ squares. Because each square represents 25 ft of travel, the car travels $72 \cdot 25 = 1800$ ft during this period of time.

4. This step asks students to count only the squares that are more than half shaded. Discuss with students why this method might work, and why it's easier than estimating a value for every partially filled square. Can students imagine a function for which this results in a bad approximation?

Q5 Answers will vary. There are approximately 44 whole squares and 7 squares that are more than half shaded, for a total of 51 squares in the left region.

Q6 Answers will vary. The car traveled approximately 1275 ft.

6. Because the squares are quite small and there are many of them, counting them one at a time is tedious. Encourage students to find ways to make the process more efficient (perhaps counting by 10's or 20's).

Q7 Answers will vary. There are approximately 320 or 330 squares in the left region. Each square now represents 4 ft of travel.

Q8 Answers will vary. A count of 325 squares corresponds to 1300 ft of travel.

Q9 This estimate is more accurate. The smaller squares allow a more accurate measurement, because they fit the curve better. You could obtain a more accurate result by making the squares still smaller.

10 There are about 597 squares between $t = 30$ s and $t = 70$ s, representing a distance of 2388 ft. The total distance traveled is about 1300 ft + 2388 ft + 1800 ft = 5488 ft.

EFINITE INTEGRALS FOR OTHER FUNCTIONS

11 Based on 14 squares, the definite integral is about 14.

12 Based on 55 squares, the definite integral is about 13.75.

13 If the function were constant, we could just find the area of the rectangle instead of counting. If the function were a linear function, we could find the area of a trapezoid. Either method would be much easier than trying to count every single square.

EXPLORE MORE

When students count the squares on page 3 of the sketch, answers will vary but should not be too far from the actual value by symbolic integration, which is 22.7 (to three significant digits).

Student answers for the suggested functions and domains will vary. Here are precise values:

$$\int_0^{\pi} \sin x\, dx = 2.0 \qquad \int_0^{\frac{3\pi}{2}} \sin x\, dx = 2.0$$

$$\int_1^2 x^2 - 2x - 1\, dx = -1\tfrac{2}{3} \qquad \int_2^3 x^2 - 2x - 1\, dx = \frac{1}{3}$$

Make sure that students have figured out that when a square is below the x-axis, they must subtract it instead of adding it.

RECTANGULAR AND TRAPEZOIDAL ACCUMULATION (PAGE 149)

Objective: Students develop a way of accumulating values that is more accurate and more efficient than counting squares.

Prerequisites: This activity will be most successful if students have completed the One Type of Integral activity. (Depending on time available, the students may be able to do both activities in a single lab period.)

Sketchpad Proficiency: Intermediate. This activity requires students to tabulate data, to use a custom tool provided with the sketch, and to use Sketchpad's Calculator.

Class Time: 25–30 minutes

Required Sketch: Trapezoidal Accumulation.gsp

DEFINITE INTEGRALS BY RECTANGLES

The first page of this sketch shows a rectangular accumulation in which you can vary the number of rectangles. The rectangles were created by a custom tool, which students will use themselves in step 9.

Q1 The sum of the areas is approximately 38.5.

Q2 The sum of the rectangles is a rather crude measurement, because the rectangles don't match the shape of the function very well. It's easy to see area under the curve that is not included in the rectangle area, and area outside the curve that is included in the rectangle area.

Q3 The largest is 38.875, and the smallest is 37.000.

Q4 The last estimate uses the smallest rectangles and so is more accurately fitted to the graph.

CONSTRUCT YOUR OWN RECTANGLES

Q5 The sum of the three rectangles is 19.38.

Q6 The sum of the three rectangles is 16.47. This is significantly closer, although by looking at the rectangles it's clear that it still overstates the true value.

DEFINITE INTEGRALS BY TRAPEZOIDS

Q7 The sum of the areas of the six trapezoids is 14.00.

Q8 The result is smaller and eliminates most of the exces of the rectangles. The trapezoids give a significantly more accurate result.

Q9 Using 12 trapezoids gives a result of 13.89, which is even more accurate.

EXTENSION

As an extension of this activity, you may want to ask students to graph their own functions and then use the **Rectangle** or **Trapezoid** tool to accumulate the area under the curve. Both tools were created to work automatically on a function labeled $f(x)$ and with a measurement labeled *width*.

IMITS WITH TABLES (PAGE 151)

bjective: Students explore the fundamental concept a limit numerically, by looking at values in a table. ıe formal definition and manipulation of delta and silon are left for a later activity.

erequisites: Students should know that 0/0 is an determinate form.

etchpad Proficiency: Beginner. Students are expected to e the Calculator, to double-click a parameter to change value, and to add data to a table. In the Explore More ction, students are asked to edit a function.

ass Time: 30–40 minutes

quired Sketch: Limits By Table.gsp

TABLE OF VALUES

Q1 The result is undefined, because the divisor is $(x - 5)$ and division by zero is undefined.

Q2 $f(4) = 3.60$, $f(5)$ is undefined, and $f(6) = 4.40$.

Q3 The parameter changes from 4.5 to 5.4 by increments of 0.1 and then returns to its original value.

5. Make certain that students select the existing table first so that the **Add Table Data** command is enabled. Also make sure students understand the effect of the "As Values Change" radio button.

Q4 The table shows an undefined value for $f(5)$. Values before $f(5)$ are increasing regularly by 0.04, and values after $f(5)$ are also increasing by the same amount. If the pattern is followed, the value of $f(5)$ is 4.000.

ET CLOSER TO THE LIMIT

Q5 The pattern indicates the same limit as before. The values of $f(x)$ now come within 0.004 of 4.000.

Q6 Yes, these values are even closer, within 0.0004 of 4.0000.

Q7 If x is within about 0.02 of 5, $f(x)$ is within 0.01 of 4. (More precisely, x must be within 0.025 of 5.)

Q8 If x is within about 0.002 of 5, $f(x)$ is within 0.001 of 4. (More precisely, x must be within 0.0025 of 5.)

EXPLORE MORE

Factoring the original function leads to

$$f(x) = 0.4\frac{(x-5)(x+5)}{(x-5)}$$

In this form, it's easy to see that at $x = 5$ the function's value becomes 0/0, making it undefined at that single point in its domain.

LIMITS WITH DELTA AND EPSILON (PAGE 153)

Objective: Students explore the formal definition of a limit and by choosing values of L (to determine the limit itself) and manipulating values of δ and ϵ (so that the definition actually makes sense).

By manipulating the sliders for δ and ϵ, students get a much clearer sense of the meanings of these two values and a clearer understanding of the formal definition of a limit.

Prerequisites: Students should complete the Limits with Tables activity before this one.

Sketchpad Proficiency: Beginner. Students are expected to double-click a parameter to change its value. In the Explore More section, students are asked to edit a function.

Class Time: 30–40 minutes

Required Sketch: Limits Epsilon Delta.gsp

FIRST FUNCTION

Q1 The point x on the axis moves to c. The red point on the graph and the connecting segments disappear. The value of the function is undefined.

Q2 When x is very close to c, the value of $f(x)$ is very close to 4, so the limit is 4.

Q3 When x is between 4.5 and 5.5, the value of the function is within ϵ of L.

Q4 A value of approximately 0.6 keeps $f(x)$ within ϵ of L.

Q5 A value of approximately 0.37 keeps $f(x)$ within ϵ of L.

Q6 Yes. No matter how small ϵ is, you can just make δ smaller as needed.

SECOND FUNCTION

Q7 Yes, a value of δ that's less than 2.0 will keep $f(x)$ within ϵ of L.

Q8 Now there's no way to choose an appropriate δ. Any value you pick allows the value of $f(x)$ to be taken from either of the two branches of the graph, both of which are outside the ϵ range. There's no other value of L that will work, because the function takes a jump from 2 to 4 at $x = 4$, and any limit can't be close to both of those values.

OTHER FUNCTIONS

Q9 When $c = 3$, there is no combination of values of L and δ that will work, because values of x slightly greater than 3 produce very large (positive) values, and values slightly greater than 3 produce very small (negative) values.

MANUALLY PROBING THE ANTIDERIVATIVE
(PAGE 155)

Objective: Students use what they know about derivatives and slopes to trace an antiderivative.

Prerequisites: Students must understand the relationship between the derivative of a function and the slopes of tangents to the function.

Sketchpad Proficiency: Intermediate. Students are expected to be able to complete simple constructions without being told what to do at each step along the way.

Class Time: 20–30 minutes. If students finish early, they could start on the next activity (Automatically Probing the Antiderivative). Most students can finish both activities in a 45- or 50-minute period.

Required Sketch: None

CREATE THE FUNCTION

1. Make sure students know how to create parameters when constructing a function.

Q1 The derivative is positive between about –5 and –1 and also for values of x greater than about 4. When the derivative is positive, the antiderivative is increasing.

BUILD THE PROBE

Q2 When the derivative is positive, the probe should point up and to the right. When the derivative is zero, the probe should be horizontal.

Q3 The value of $f(x_p)$ is equal to the slope of the antiderivative at the same x-value.

Q4 If you use 1 as the value of Δx, the value of Δy is equal to the value of $f(x)$.

Q5 The line points up to the right when x is between approximately -5.15 and -0.95, and also when it's greater than about 4.1. These are the same values from Q1. The line is horizontal at approximately $x = -5.15$, $x = -0.95$, and $x = 4.1$. The line points down to the right for values of x less than approximately -5.15, and for values between approximately $x = -0.95$ and $x = 4.1$.

The explanation lies in the value of the function $f(x)$, because this value determines the slope of the line. Whenever $f(x) > 0$, the line points up; when $f(x) = 0$, the line is horizontal, and when $f(x) < 0$, the line points down.

USE THE PROBE

Q6 The traces indicate the direction of a possible tangent line at each spot on the screen. Students are creating a slope field here; it's up to you to decide whether to draw special attention to this phenomenon and name it at this time.

11. In this step, students are using a form of Euler's method: Calculate the direction in which to move (using the probe), move a short distance in the calculated direction, and then recalculate. Though there's no need to name the method, you should call attention to this technique for approximating solutions when only local information is available.

Q7 The three traces should all seem to have roughly the same shape, though they are displaced vertically from each other.

AUTOMATICALLY PROBING THE ANTIDERIVATIVE (PAGE 157)

Objective: Students continue the previous activity by automating the tracing process, using a movement button and using the **Iterate** command.

Prerequisites: Students must have completed the Manually Probing the Antiderivative activity and must have saved their sketch from that activity.

Sketchpad Proficiency: Intermediate. Students will be expected to be able to complete simple constructions without being told what to do at each step along the way.

Class Time: 20–30 minutes. If students finish early, the Explore More section suggests several extensions.

Required Sketch: Students must have completed and saved the sketch for the Manually Probing the Antiderivative activity.

TRACE WITH A MOVEMENT BUTTON

Q1 The automatic trace will be smoother than the manual traces but should have approximately the same shape.

Q2 The five traces all have the same shape. Because of the greater accuracy of automated tracing, the fact that the shapes are identical should be even clearer than when the students did the traces manually.

TRACE BY ITERATION

Q3 The iterated shape is similar to the one produced by the movement button but is not as smooth or accurate. This is because the iteration is using much larger steps than the movement button used.

Q4 When the value of h becomes smaller, the segments become shorter. This smooths the jagged nature of the original iteration but also shortens the iteration itself so that it shows very little of the solution.

Q5 With shorter steps at each stage of the iteration, the result is now extremely smooth and accurate. Because the direction is evaluated so frequently, there's much less opportunity for errors to accumulate.

Q6 The shorter the steps, the smaller the errors that distort the final shape. To get a completely accurate shape, we'd need to find the limit of the shape as the length of the steps approaches zero (and the number of steps increases without limit).

Q7 As you move P up and down, the shape doesn't change at all.

Q8 There are an infinite number of solutions, because for any given x-value, each possible y-value generates a different solution. Explanations vary. In terms of the probe, the solution you get depends on the y-value at which you start the probe. Stepping away from the probe, a more general answer is that if it's the slope that matters, then you can translate the original function up or down without affecting its derivative.

EXPLORE MORE

Q9 Different values of the parameters result in different shapes for the antiderivative. The important thing is that students recognize that the antiderivative is increasing where the original function has a positive value, and that it's decreasing where the original function's value is negative.

Q10 The functions students observe will vary. Examining various sine functions should help students consolidate their understanding of the relationship between the value of the function and the slope of the antiderivative.

Comment Form

Please take a moment to provide us with feedback about this book. We are eager to read any comments or suggestions you may have. Once you've filled out this form, simply fold it along the dotted lines and drop it in the mail. We'll pay the postage. Thank you!

Your Name ______________________________

School ______________________________

School Address ______________________________

City/State/Zip ______________________________

Phone ______________________________

Book Title ______________________________

Please list any comments you have about this book.

____

____

____

____

____

____

____

____

____

____

Do you have any suggestions for improving the student or teacher material?

____

____

____

____

____

____

____

To request a catalog, or place an order, call us toll free at 800-995-MATH, or send a fax to 800-541-2242.
For more information, visit Key's website at www.keypress.com.

Please detach page, fold on lines and tape edge.